Communications
in Computer and Information Science　　938

Commenced Publication in 2007
Founding and Former Series Editors:
Phoebe Chen, Alfredo Cuzzocrea, Xiaoyong Du, Orhun Kara, Ting Liu,
Dominik Ślęzak, and Xiaokang Yang

More information about this series at http://www.springer.com/series/7899

Safaa O. Al-mamory · Jwan K. Alwan
Ali D. Hussein (Eds.)

New Trends in Information and Communications Technology Applications

Third International Conference, NTICT 2018
Baghdad, Iraq, October 2–4, 2018
Proceedings

Editors
Safaa O. Al-mamory 🆔
University of Information Technology
and Communications
Baghdad
Iraq

Ali D. Hussein
University of Information Technology
and Communications
Baghdad
Iraq

Jwan K. Alwan
University of Information Technology
and Communications
Baghdad
Iraq

ISSN 1865-0929   ISSN 1865-0937 (electronic)
Communications in Computer and Information Science
ISBN 978-3-030-01652-4   ISBN 978-3-030-01653-1 (eBook)
https://doi.org/10.1007/978-3-030-01653-1

Library of Congress Control Number: 2018956292

This Springer imprint is published by the registered company Springer Nature Switzerland AG
The registered company address is: Gewerbestrasse 11, 6330 Cham, Switzerland

Preface

The Third International Conference on New Trends in Information and Communications Technology Applications (NTICT 2018), held in Baghdad, Iraq, during October 2–4, 2018, was hosted and organized by the University of Information Technology and Communications. NTICT is an international conference focusing on specific topics in computer networks and communications, artificial intelligence, and machine learning. NTICT 2018 was the first conference in the series to have its proceedings be published in *Communications in Computer and Information Science* (CCIS) by Springer.

NTICT 2018 aimed at providing a meeting for an advanced discussion of evolving applications in artificial intelligence, machine learning, computer networks, and communications. The conference brought both young researchers and senior experts together to share novel findings and practical experiences in the aforementioned fields. The NTICT 2018 conference enhanced the scientific research on this subject in Iraq and the rest of the world.

The call for papers resulted in a total of 88 submissions from around the world. Every submission was assigned to at least three members of the Program Committee for review. The Program Committee decided to accept 18 long papers, which were arranged into seven sessions, resulting in a strong program with an acceptance rate of 21%. The accepted papers were distributed in seven tracks (as specified in the call for paper) Computer Networks, System and Network Security, Machine Learning, Computer Vision, E-learning, Intelligent Control System, and Communication Applications.

We would like to thank all who contributed to the success of this conference, in particular the members of the Program Committee (and the additional reviewers) for carefully reviewing the contributions and selecting a high-quality program. The reviewer's efforts to submit the review reports within the specified period are greatly appreciated. Their comments were very helpful in the selection process. Furthermore, we would like to convey our gratitude to the keynote speakers for their excellent presentations. The word of thanks is extended to all authors who submitted their papers and for letting us evaluate their work. The submitted papers were managed using the Open Conference System (OCS); thanks to Volha Shaparava for the continuous support. Finally, we thank Prof. Aliaksandr Birukou, Leonie Kunz, and Tamara Welschot for making these proceedings possible.

We hope that all participants enjoyed a successful conference, made a lot of new contacts, engaged in fruitful discussions, and had a pleasant stay in Baghdad.

October 2018

Safaa O. Al-mamory
Abbas Fadhil Aljuboori
Mouayad Sahib
George S. Oreku

Organization

General Chair

Abbas Muhsin
Albakry

President of University of Information Technology
and Communications, Iraq

Program Chairs

Safaa O. Al-mamory

University of Information Technology and communications,
Iraq

Mouayad Sahib

University of Information Technology and communications,
Iraq

George S. Oreku

Open University of Tanzania (OUT) – North West
University, South Africa

Abbas Fadhil
Aljuboori

University of Information Technology and communications,
Iraq

Guest Editors

Hilal Mohammed
Yousif Albayatti

Director of Quality Assurance and Accreditation Centre,
Bahrain

Alaa Hussein
al-Hammami

Princess Sumaya University for Technology, Jordan

Mudafar Kadhim Ati

Abu Dhabi University, UAE

Steering Committee

Jane J. Stephan

University of Information Technology and Communications,
Iraq

Zaidon A.
Abdulkariem

University of Information Technology and Communications,
Iraq

Mouayad Sahib

University of Information Technology and Communications,
Iraq

Sinan Naji

University of Information Technology and Communications,
Iraq

Buthaina Fahran Abd

University of Information Technology and Communications,
Iraq

Firas Abdul Hameed
Abdul Latef

Ibn Al_Haitham College, University of Baghdad, Iraq

Ali Hassan Tarish

University of Information Technology and Communications,
Iraq

Intisar Shedeed jlaib	University of Information Technology and Communications, Iraq
Inaam Rekan Hassan	University of Information Technology and Communications, Iraq
Atheer Akram Abd AlRazaq	University of Information Technology and Communications, Iraq
Ali Abdulhadi Jasim	University of Information Technology and Communications, Iraq
Mohammed Q. Mohammed	University of Information Technology and Communications, Iraq
Forat Kadhim Challoob	University of Information Technology and Communications, Iraq
Ahmed A. Hashim	University of Information Technology and Communications, Iraq
Samer A. Kahioosh	University of Information Technology and Communications, Iraq
Ali Dakhel	University of Information Technology and Communications, Iraq
Samer Mohammed Ali	University of Information Technology and Communications, Iraq

International Scientific Committee

Abdel-Badeeh Salem	Ain Shames University, Egypt
Alaa Al-Shammery	University of Information Technology and Communications, Iraq
Alaa Hussein al-Hammami	Princess Sumaya University for Science and Technology, Jordan
Ali Abdulhadi Jasim	University of Information Technology and Communications, Iraq
Ali Al-Sherbaz	The University of Northampton, UK
Ali H. Al-Timemy	University of Baghdad, Iraq
Dennis Lupiana	Institute of Finance Management, Dar es Salaam, Tanzania
Eman Alshamery	University of Babylon, Iraq
Eva Volna	University of Ostrava, Czech Republic
George S. Oreku	Open University of Tanzania/North West University, South Africa
Ghaidaa Al-sultany	University of Babylon, College of Information Technology, Iraq
Hasan Fleyeh	Dalarna University, Sweden
Hilal Mohammed	Applied Science University, Bahrain
Hoshang Kolivand	Liverpool John Moores University, UK
Mahdi Nsaif Jasim	University of Information Technology and Communications, Iraq
Malleswara Talla	Concordia University, Canada
Mohammad Shojafar	Sapienza University of Rome, Italy

Contents

Computer Networks

Clustering an Unstructured P2P Networks Using a Termite Hill Building Model

Hazim Aburagheef[1]([✉]) [iD] and Safaa O. Al-mamory[2]([✉]) [iD]

[1] College of Information Technology, University of Babylon, Hillah, Iraq
hazim@itnet.uobabylon.edu.iq
[2] College of Business Informatics,
University of Information Technology and Communications, Baghdad, Iraq
salmamory@uoitc.edu.iq

Abstract. Super-peer (SP) architecture is proposed to improve the quality of service (QoS) of peer to peer (P2P) networks. P2P networks is divided into sets of homogeneous sub-groups representing the number of SPs. Designing SP networks for file sharing has several issues like the specifying best number of SPs, selection of SPs, and suitable ordinary peers for each SP. In this paper, we propose a simple method to achieve self-organization of peers in dynamic environment to enhance QoS. Termite hill building model is used for clustering an unstructured P2P network by employing Jaccard measure to compute peers' interest similarity. This method consists of four steps which are initialization, separation, colony building, and post processing. Both the separation and colony building steps are the backbone of the method. The experimental results on a simulated network with 10000 nodes show about 99% as accuracy.

Keywords: Swarm intelligence · P2P clustering · Termite hill building

1 Introduction

P2P network is a distributed network includes distributed computing, content sharing, collaborative systems, and platform services. P2P content sharing is a virtual overlay consisting of set of nodes connected with each other [1]. These networks can be classified into structured, unstructured, and hybrid networks [2]. Resource discovery in decentralized environments provides important challenges caused by the scalability, dynamicity, and heterogeneity of these environments. P2P systems can be very large as well as the size of information shared in these systems. Additionally, these systems are extremely dynamic with peers randomly joining, leaving, and failing [3].

In this paper, the main problem to focus on is resource discovery in unstructured P2P networks. In addition, the problem of superpeers selection and the best number of nodes (peers) each superpeer can serve are examined. Clustering provides a method to structure a P2P system by building a virtual overlay network from an unstructured P2P network. The purpose of clustering is to add some controlling on networks that are varied and changed with time. There are different methods used in P2P network clustering, such as Interest-based clustering [4], content-based clustering [5], semantic-

© Springer Nature Switzerland AG 2018
S. O. Al-mamory et al. (Eds.): NTICT 2018, CCIS 938, pp. 3–20, 2018.
https://doi.org/10.1007/978-3-030-01653-1_1

based clustering [6], topology-based clustering [7], density-based clustering [8], k-mean clustering [9], and Growing Neural Gas clustering [10].

The suggested solution exploits the interest-based similarity by grouping all nodes having same interest in single layer and then dividing each layer into set of clusters. In other words, we propose a simple method, using the termite hill building model, to solve the super peers problems mentioned above. Any node having the maximum number of pheromones can be considered as a super peer. The simulation accuracy on a network containing 10000 nodes is about 99%.

The remainder of this paper is organized as follows. Section 2 presents a review of related work. In Sect. 3 we explain the termite hill building model. Section 4 presents in detail the proposed method. Section 5 presents the experimental results. Finally Sect. 6 draws conclusions.

2 Related Work

Clustering is important topic in P2P network and can be defined as dividing P2P network into similar sub networks. Sharing file contents is a principle element in P2P network. This section reviews of P2P network clustering techniques for both ordinary and self-organization (bio-inspired and non-bio inspired methods).

Atul et al. [11] used economic Schelling model in which the space domain is modeled as a two dimensional grid of cells having label color. Two clusters is first been built using selfless clustering algorithm which contains peers finding similar neighbors. Lakshmish et al. [12] uses node weight depending on the degree of a given node. Clustering begin with selecting originate nodes in graph then each originate nodes send re-circulate message into graph with limited TTL. Each node builds a table of weights for each message arrived from originate.

Saurabh et al. [13] uses replica density of file. The cluster classified into two classes which are higher replica density (if clusters' files are more popular than other clusters) and lower replica density (if clusters' files are not popular). These clusters are either heavy clusters or poor clusters depending on replica density skew. The authors used replica density in forwarding query to the clusters having high replica density when used random walk searching method.

Ayyasamy et al. [14] exploited QoS aware topology depending on access pattern. Peers are belonging into strong or weak clusters depend on their weight vector. Weight vector includes available capacity, speed of CPU, size of memory, and access latency parameters. Replica placement algorithm classified content of each peer into class one (most frequently accessed contents) and class two (less frequently accessed contents). All content of class one are replicated in strong clusters and all content of class two are group in weak clusters. Routing is executed hierarchically through sending the query only to the strong clusters.

Ma et al. [7] present a Group-Average Agglomerative Clustering Topology (GAAC) algorithm. Each node is a sub cluster of its above cluster except the node root. GAAC base on the similarity function in both individual and grouping. Individual Similarity is a function of three-dimension vector including successful communication, amount of communication data, and group similarity.

Dumitrescu et al. [10] employed a neural network model called Growing Neural Gas (GNG). It is neural network model talented to increasingly learn the important topology relations through a given set of input vectors by applying of Hebb like training simple rule. Peer sends out a search super-peer message and waits for the response of two super-peers. It only accepts first two answers. The round trip time of the match messages are recorded, after arrival of response find superpeer, peer send update message, which consist of four tuples, includes two superpeers and two round trips time.

Michalis et al. [15] used Semantic Overlay Network (SON). SON is group together peers that contain similar contents. First, use classification to generate groups of files, and following peers. Classification is supervised methods; clustering algorithms are suitable for SON generation. First step is building feature vector consisting of two values: first feature (word) and second value is weight. Each cluster is summarized into a feature vector. Network is divided into equal zones and select initiate peer for each zone. Initiators are uniformly distributed in the network and a suitable number of initiators proportionate to the total size of network. Super-clusters are built from merge similar clusters, neighbor zones set are grouped into generate a super-zone, and neighbor super-zone are merge into larger super-zone.

Tirado et al. [16] used the self-organizing affinity P2P for large scale networks. Their system raised replication of content and reduced the search latency time. It suggests that each peer automatically joins or leaves a cluster depending on its contents. Each peer joins one primary cluster and more than one secondary clusters, depending on its contents. An affinity matrix is applied to compute the semantic closeness between clusters. A user may have variety interests.

Golnaz et al. [17] used improved reinforcement learning-based approach depending on the basic results of decision theory. Each peer used concurrent learners to monitor the behavior of other participants in their neighborhood and is learning from these monitoring how to keep up with the environment so that constantly tracks the shifting behavior. Particle swarm optimization is employed based on a stochastic evolutionary algorithm.

Ebrahimi et al. [18] used ant-based clustering algorithms as a new peer clustering algorithm for P2P database systems. They have modified ant-based clustering algorithm to group peers having similar contents which makes the queries directed to the suitable cluster and hence to be effectively answered. Ants move, as they proposed, in two-dimensional spaces and compute the picking up probability depending on the specific criteria for peers they meet with. If possible, each ant picks up a peer and puts it in another location while computing the similarity of peers. The authors claim that the proposed algorithm has high scalability against increasing problem size while the number of peers in their experiments was 2000 peers.

Meng [19] improved the data availability in P2P data storage systems under churn calculating online probabilities of peers in several intervals. Then, they group the peers having the complementary online patterns in order to create the generalized peers with high online probability in each time interval. A generalized peer consists of several member peers with complementary online patterns, which means if a data object is stored on all the member peers of a generalized peer, then its availability can be guaranteed in any time interval under churn.

Meng et al. [20] discussed topology constructing of an unstructured P2P networks taking into consideration the free riders and retrieval efficiency. These two parameters are considered to detail the topological construction and adjustment mechanisms. Their goal was to reduce the percentage of free riders and improve the retrieval efficiency in unstructured P2P networks.

The proposed system is different from these systems in several aspects. Most of these aspects are presented in Table 1 containing main information such as similarity measurement, accuracy of clustering, purpose of each clustering method, and other parameters.

Table 1. General comparison between different clustering methods

Source	Type	Scale	Similarity measurement	#Clusters	Accuracy	Aim
[11]	Economic (Schelling)	5000	Percentage of neighbors	Begin with fixed then reduce	Scale converge	Clustering
[12]	Connection	5000	Weight of message which reduce by function of degree	Fixed	Scale converge	QoS
[13]	Popularity	–	–	Fixed and equal size	–	Improve searching
[14]	Replication	1000	–	Fixed	–	QoS
[7]	Topology	800	Distance of vectors	Fix no. then merge between similar clusters	Group-average similarity	Routing efficiency
[10]	Super-peer	<500	Vector of age to every edge and an error counts	Dynamic	Average distance	QoS
[15]	Semantic	10000	Cosine similarity	Fixed at beginning	F-measure	Improve searching
[16]	Self-organized	5000	Content categories	Fixed and equal to categories	Affinity matrix	Improving service
[17]	Self-organized		Corporation ratio	Q-learning	decision-analytic	QoS
Proposed system	Self-organized	>10000	Jaccard	Dynamic	Interest similarity	QoS

3 Termite Hill Building Model

The behavior of termites in hill building is an example of swarm intelligence. In hill building colony, termites want to gather pebbles spread over an area into one place in order to build a hill. Termites are individuals act independently and move only on the basis of the observed local pheromone concentration [21]. Each termite follows four rules depending on pheromone states. Firstly, if no pheromone exists, a termite moves uniformly randomly in any direction. Each termite may carry only one pebble at a time. If a termite is not carrying a pebble and it meet by chance one, the termite will pick it up. If a termite is carrying a pebble and it encounters one, the termite will put the pebble down. The pebble will be infused with pheromone which then evaporates and creates a concentration for others to follow. With these rules, a group of termites can collect dispersed pebbles into one place [22].

Secondly, if a pheromone exist (positive feedback), positive feedback is represented by a termite's attraction towards the pheromone concentration. The termite is biased to add more pebbles to large piles. The larger the pile, mean the more pheromone it is likely to have, and more pebbles will be moved to it. The greater the bias to the hill, the faster termites is likely to arrive, further increasing the pheromone content of the hill.

Thirdly, the pheromone evaporates (negative feedback), it consequently weakens and lessens the resulting concentration. A diminished concentration will attract fewer termites as they will be less likely to move in its direction according to the concentration. While this may seem detrimental to the task of collecting all pebbles into one pile, it is in fact essential. As the task begins, several small piles will emerge. Those piles that are able to attract more termites will grow faster. As pheromone decays on lesser piles, termites will be less likely to visit them again, thus preventing them from growing. Once all of the pebbles in the small piles have been picked that pile will cease to exist and can never grow again; larger piles will grow instead. Negative feedback, in the form of pheromone decay, helps large piles grow by preventing small piles from continuing to attract termites [23].

Finally, if not enough termites exist then the pheromone would decay before any more pebbles could be added to a pile. Termites would continue their random walk without forming any significant piles. Where and when piles are created or destroyed is determined entirely by chance since each termite makes independent and probabilistic decisions; termites use pheromone to coordinate their activities [24].

The using of termite algorithm in the proposed model is explained in more details below; however it may be described simply as follows. Each node in the network is represented as a pebble. A termite needs pebble to build a pile (colony). A termite searches in environment on this pebble and carry it to first pebble detection in path. Each termite searches on different types of pebble depending on the source pile (source node).

Each class of interest can be represented as independent hill (pile) and each hill can be represented as colony. Each colony has a single queen. Queen can born large number of forages called termites. These termites work to build a colony. Colony is consisted of all nodes having the same interest. Any colony has a number of heads and each head represents a cluster.

Each node in the network has a specific pheromone table. As termite move through the network on links between nodes, they are biased towards the pheromone to find nearest pebble (similar of source node). When it finds a pebble it carries the pebble and put it on the pile (source node). Any termite keeps searching for pebbles in order to building a big pile.

3.1   The Pheromone Table

In order to manage the pheromone in the network, each node maintains a table recording the pheromone on each neighbor link. Each termite has a pheromone scent. The table may be visualized as a matrix with neighbor nodes listed contains pheromone concentration the side. Termite selects the link having maximum pheromone concentration. Initial value of the table is computed by Eq. (1).

$$p(i) = \frac{\frac{Q}{T(i)}}{\sum_{j=0}^{s} \frac{Q}{T(j)}} \tag{1}$$

Where T(i) is the delay time between nodes i and j, S is a set of neighbors, Q is a constant integer quantity. Initial value applied as a function to inverse delay time between two neighbor nodes.

3.2   Pheromone Update

Termite follows this pheromone concentration while laying pheromone for their source on the links that they traverse. The amount of pheromone deposited by a termite on a link due Eq. (2). Using this method of pheromone updating, consistent pheromone trails are built through the network.

When a termite is arrived at a node n from p, the pheromone (Ph) is updated in the pheromone table according to Eq. (2) with a constant amount of pheromone. When a termite passed each node, the node should update its pheromone table in this way.

$$p_i(t+1) = p(t) + \rho\left(\frac{Q}{T_j}\right) \tag{2}$$

where p is the previous pheromone value, T is delay time between nodes i and j, Q is constant integer quantity.

3.3   Pheromone Evaporation

To account for pheromone decay, each value in the pheromone table is periodically multiplied by a decay factor. A high decay rate will quickly reduce the amount of remaining pheromone, while a low value will degrade the pheromone slowly. Equation (3) describes pheromone decay.

$$p_i(t+1) = p(t) - K(p(t)) \tag{3}$$

where K is decay factor, $0 < K < 1$. The decay factor value is selected depending on the state of the system. K is selected to have high value if it needs speed to arrive at a steady state, and vice versa. If all of the pheromone for a particular node decays, then the corresponding row is removed from the pheromone table. Removal of an entry from the pheromone table indicates that no termite has been received from that node in quite some time.

4 Proposed System

Clustering operation is an important step and all subsequent steps depending on it. The purpose of this phase is separating P2P network into set of layers and each layer contains the same interest peers. Each layer is divided into set of clusters. We used termite ant hill building model in interest-based clustering. Clustering operation consists of four steps which are initialization, separation, colony building, and post processing.

Initialization steps (the first three lines in Algorithm 1) include set of operations like graph construction using Poisson distribution, file sharing distribution, delay time generation between each neighbor (here delay generation due to simulator tool), and set initial value for other parameters such as size of network, mean of neighbors (Poisson mean), delay time (range), number of cycles.

The purpose of the separation step is to separate network into independent sub groups (layers). It has two levels, the first level is the initial level where only each node tries to find the nearest similar node and initial group (one only). This level consists of two algorithms: the initialization and separation algorithm which is described in Algorithm (1), and the find similar node algorithm depicting in Algorithm (2).

Algorithm (1): Initialization and separation step algorithm

```
Input: #Cycles, termite live time (Tt), Poisson dis-
tributed mean, files sharing, Min and Max delay
time.
Output: set of similar groups.
Begin
Graph construction
File sharing distribution
Path time generation
Repeat
        For each  peer has not similar neighbor do
            find similar node ( )
        End for
        Update termite live time
Until #cycles exceeds
End.
```

In the find similar node algorithm, a termite scans for all nodes that not exceeds termite live time and is arrived from this node. Termite ant always selects the path having minimum time delay in each hop using best neighbor selection algorithm which depends on the probability function described in termite hill building model (Eq. (1)), and select neighbor which has maximum pheromone value. When termite is arrived into live time and not finding similar node, termite go back and select second best neighbor and so on until scan all nodes. In each hop, termite live time is reduced by time of path between termite node and best neighbor and when back termite live time is increased by time of path.

Algorithm (2): find similar node algorithm

```
Input: Node (N), termite live time (Tt), set of
neighbor (M)
Output: set of groups
Begin
Create termite ant
While (set of neighbors(N) is not empty) do
        S = select best neighbor (N)
        R = all nodes arrived from node (S) and
not exceed  termite live time
        If test similar (N,R) then
            Return  group(N, nearest one)
        End if
End while
Return null
End
```

Algorithm (2) only finds one similar peer neighbor and initial group. When a termite finds similar peer, a new virtual connection with group is created; then, compute pheromone for new peer and edge of group through computing delay time between them. The next cycles are only applied on peers haven't similar neighbors (group); the termite live time is updated with small delta time.

The second level of separation step uses a number of cycles. It uses these groups to extend to similar groups and continues until it finds all the similar groups, and they bind them with each other in one sub group (layer). This step consists of different algorithms: the separation algorithm described in Algorithm (3).

Algorithm (3): Second level separation.

```
Input: graph, set of groups, delay
time (Td), #Cycles
Output: set of layers
Begin
Repeat
  For each group in graph  do
    For each edge of group do
      test node Similarity ( )
    End for
  End for
Until #Cycles exceed
End.
```

In Algorithm (3), a termite begins with any node (start node) belong to a group. It travels from one node to another using pheromone concentration; the neighbor node selection depends on maximum pheromone value. Termite must arrive to all edges in group and begin with nearest edge. When termite arrives into edge it begins to search for similar neighbor group using find similar node algorithm. Termite updates pheromone for each node of path (from source node and new node detection) if it finds a new similar node. An evaporation operation is done at the end of each cycle.

The colony building step consists of a number of operations. The first level is to select a set of nodes where each one has a pheromone value greater than threshold and represents a queen termite. The pheromone value is computed according to Eq. (4).

$$Ph_n = \sum_{i=0}^{s} Ph_i \qquad (4)$$

where Ph is pheromone value of node (n), S is the set of neighbors

The second level is building a colony for each member in set of nodes resulted from first level. Each iteration generates only one colony building with evaporation in each cycle in order to separate each colony from others. Colony building steps is described in Algorithm (4). Delay time is the maximum path length from center node (super-peer) and it equals to termite live time. Termite uses pheromone concentration with delay time, to pass through all nodes which distance is not exceed delay time and it begins with short path. Termite will label each node pass through it and updates pheromone value, termite uses this label to prevent a node from belong to other colonies.

The post-processing step is the final step to treat nodes which doesn't belong to any colony after doing the clustering. Essentially each node does not belong to any colony is free; these free nodes will belong to nearest similar colony. Termite live time is an important variable in this step.

We designed these steps to reduce the cost by pairing each node with their nearest neighbors. In each new cycle we added delta time to the original delay time in order to pairing any node with nearest nodes. In the first level of separation step, the created termite number is equal to the number of nodes with smallest termite live time. In the next cycle, the number of created termites is equal to the number of nodes not paired (grouped); this process will continuous with increasing termite living time and decreasing the number of termites. Therefore, there is equilibrium between the number of termites and termite living time. In second level of separation process, the number of created termites is equal to the number of groups which produced in the previous level. In each next cycle, the number of termites is equal to the number of groups and is reduced until arriving to the number of classes.

Algorithm (4): Colony building algorithm.

```
Input: set of layers, Pheromone thresh-
old (Ph), delay time (Td)
Output: set of clusters
Begin
Compute pheromone for each node by
Equation (4)
Repeat
  N = Maximum pheromone in layer
  Colony building (N, Td)
  Reset Pheromone (N)
    Evaporation
Until node pheromone less than Ph
End
```

The dynamic environment is one of the most important attributes of P2P networks. Any system looking forward to improve P2P networks must take into account this attribute. The proposed system treats this state with different methods depending on the state of the system. The dynamic environment in the separation step has no effect on the system because when a node state is updated, departure or joining, it can be detected in the next cycle. Therefore, the effect will be only on the network size (reduction or increasing, respectively). To prevent losing the clusters when queen node is lost, another node (i.e. the neighbor of the queen node) will be nominated to be the queen.

5 Results

Clustering experiment has been applied with the aim of discovering the most important variables and their best values, and the effect of each variable on the other variables. Each step of the proposed system has an important variable. Proposed system consists of two main steps and one secondary step. Main steps are separation step (layer separation) and colony building. Secondary step is post processing step. The main goal of separation step is to separate network to sets of layers equal to classes' numbers. Colony building step aims to clustering each layer. Peers not belonging to any cluster are treated by post processing step. PeerSim [25] simulator is used with event cycle. The random graph (Poisson distribution) topology used to generate the unstructured P2P model through degree estimation for each node. For the object placement and query strategies, uniform and Zipf distributions are used in the experiments.

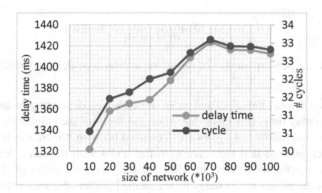

Fig. 1. Relation between network size verse delay time (left) and #cycles (right)

The first experiment is to study the separation step in more details including the effect of each important variables and relation between them. The purpose of the experiment is to know the best values of cycles, delay time, mean of neighbors, and number of classes. When we know the value of any one of these values, we can estimate the values of the others. The relation between the network size verse both the number of cycles and the delay time is depicted in Fig. 1. It is clear that number of cycles and delay time are increased when the network size increased and only one cycle can be used when the initial delay time is larger than (1500 ms) but in this state the node does not select the nearest similar neighbors because the algorithm depends on the inverse (select minimum) greedy algorithm. Therefore, we applied the initial level of the separation step with a multi cycle to pair with the nearest similar neighbor node. Both curves in Fig. 1 are degraded when the network size is larger than 70000 nodes. This is because the access to any node (not belonging to any layer) from more than one path becomes possible. Moreover, the number of nodes belonging to the same layer become large and then increases the probability of access within lower number of cycles.

In Fig. 2, the effect of the number of classes on both number of cycles and delay time is clear. Uniform random generation is used for this experiment. Each node's computed similarity depends on the shared contents. The number of cycles and delay time variables are increased when the number of classes increased. The number of layers increases when the number of classes increased. Therefore, we need more cycles and delay time to separate these layers.

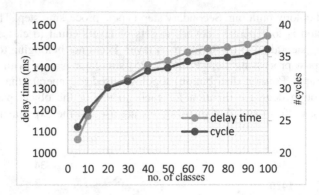

Fig. 2. Relation between number of classes verse delay time (left) and #cycles (right)

In Fig. 3, we display the effect and mean of neighbors, depending on the number of cycles and the delay time. It is clear that when the mean of the neighbors is increased, both delay time and number of cycles are decreased because it increases the number of choices for each node to find the nearest similar nodes with the same delay time and that affects the number of cycles. In addition, the increased mean of the neighbors always reduces the values of the delay time and the number of cycles regardless of the network size.

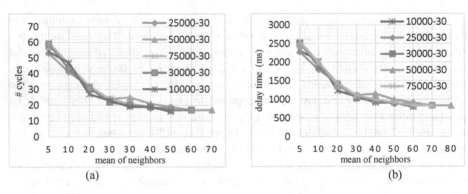

(a)   (b)

Fig. 3. Mean of neighbors effect on (a) number of cycles, (b) delay time, when network size (10000, 25000, 30000, 50000, and 75000) nodes and number of classes is 30

The purpose of the separation experiment step is to know the value of network variables. For example when we work on a network size with 50000 nodes and one variable value is known (delay time, number of cycles, class domain, or mean of neighbors) such mean of neighbors is (30). From Fig. 3, it could be concluded that the best value of the number of cycles and delay time are equal to (25) and (1200 ms) respectively.

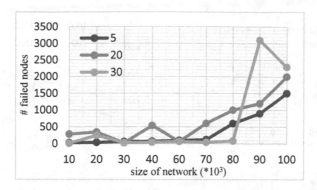

Fig. 4. Fail node in separation step, when number of classes are (5, 20, and 30)

When the network grows the accuracy is decreased, and to improve the accuracy either the delay time or number of cycles must be increased. The increasing in network size leads to rise the percent of node failure. An experiment is conducted with 1200 ms as delay time, 25 as number of cycles, and three different numbers of classes. In Fig. 4, the number of failed nodes is high when the network size and number of classes are increased. It can be noted from the figure that the number of failed nodes is rising quickly when the network size is 80000 and more. The reason of the failure is that the delay time (1200 ms) is not enough to access this big number of nodes.

Another experiment is applied to colony building and the post processing steps to study the important attributes of threshold, delay time, and network size. Threshold is used to decide the number of clusters; delay time is used to determine the size of each cluster. This experiment shows the effect of each attribute on the accuracy of the system, and number and size of clusters.

The relationship between the size of the network and the threshold with effect on both cluster number and size of number can be seen in Fig. 5. Here number of classes is fixed and threshold value is changing in order to select best value of the threshold. Both mean number of clusters and mean size of clusters are measured in two states. In first state, it measured with different threshold and different delay time values; the results are depicted in Fig. 5(a, c). In Fig. 5(b, d), the average size and number of clusters are computed for different threshold values in order to know the effect of delay time on both size and number of clusters. It is clear from these figures that both number and size of clusters increases when network size increases and vice versa. Both threshold and delay time affect the number and size of the clusters.

Threshold and delay time have an effect on the resulted number of clusters. Figure 6 shows the effect of each feature on the number and size of the clusters. It shows that the average number of clusters is affected by threshold while the delay time affects both the number and size of clusters. Delay time has more effect than threshold because when the delay time value increases.

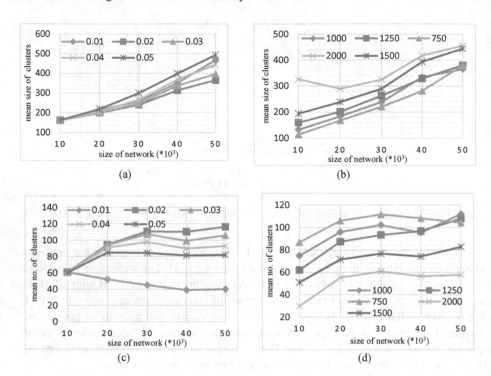

Fig. 5. Threshold and delay time effect on cluster size (a, b) and number of clusters (c, d); when threshold value (0.01, 0.02, 0.03, 0.04, and 0.05) and delay time are (750, 1000, 1250, 1500, and 2000) ms

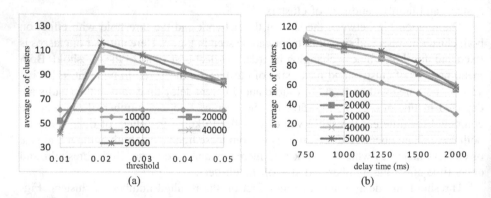

Fig. 6. Effect of (a) threshold on number of clusters (b) delay time on the number of clusters, when network size (10000, 20000, 30000, 40000, and 50000) nodes

The number of nodes belonging to the cluster increases and then the number of nodes is declined gradually with increased iterations. Furthermore, the lower the threshold value is, the higher the probability of containing nodes having pheromone greater than threshold and vice versa. Therefore, when the threshold value is 0.02, we

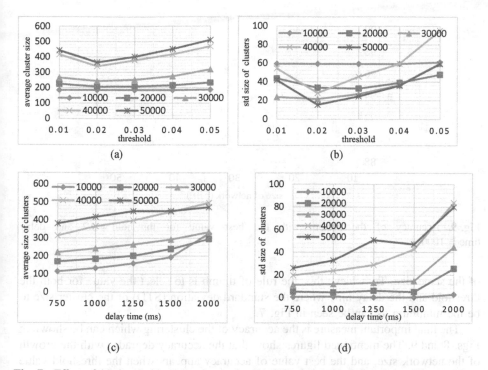

Fig. 7. Effect of (a) threshold on the cluster size (b) STD threshold on the cluster size (c) delay time on the cluster size (d) STD delay time on the cluster size, when network size (10000, 20000, 30000, 40000, and 50000) nodes.

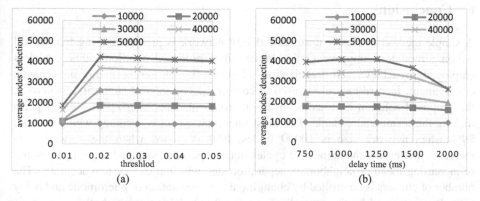

Fig. 8. Average node detection (a) threshold (b) delay time, when network size (10000, 20000, 30000, 40000, and 50000) nodes

get the better results with respect to detected nodes, and the number of homogeneous clusters' sizes. This explains the resulted knee in Fig. 6(a).

The estimation of threshold and delay time best values depends on the goodness of the cluster condition. It is important for good clusters to produce a number of clusters

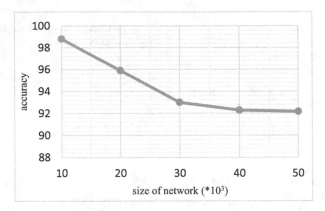

Fig. 9. Accuracy of the system with the best value for threshold = 0.2 and delay time = 1000 ms

of the same size. To achieve this, the rule of thumb is to select the value for both the threshold and the delay time where the standard deviation (STD) of the cluster size to be the minimum as can be seen in Fig. 7.

The final important measure is the accuracy of the clustering which can be shown in Figs. 8 and 9. The mentioned figures show that the accuracy degrades with the growth of the network size, and the best value of accuracy appears when the threshold value and the delay time are (0.02) of the quality factor and (1000 ms) respectively. The accuracy of the proposed system is about 92% in more than 50000 nodes.

6   Conclusions

A simple method to cluster unstructured P2P network is presented using termite hill building model to improve QoS. Unlike other proposed solutions, this model is employed to achieve self-organization of peers (by clustering them) in dynamic environment. Jaccard measure is used to compute peers' interest similarity. Moreover, the proposed solution is tested with an environment containing up to 100000 peers making the results more stable. The most important aspect is accuracy which is about 99% when network size is 10000 nodes. It is degraded when the network size increases. Increasing the number of cycles in the separation step and in post processing or increasing mean of neighbors in separation step will improve system accuracy. The number of clusters is controlled by changing the concentration of pheromone and delay time. It is increased by the controlling the threshold, delay time, or both.

As a future work, the proposed system can be used to enhance QoS which includes searching improving by reducing both number of message and mean value of TTL. In addition, the throughput of system in unstructured P2P networks can be improved through load fairly distribution.

References

1. Schollmeier, R.: A definition of peer-to-peer networking for the classification of peer-to-peer architectures and applications. In: 1st International Conference on Peer-to-Peer Computing, pp. 101–102 (2001)
2. Shen, X., Yu, H., Buford, J., Akon, M.: Handbook of Peer-to-Peer Networking. Springer, Heidelberg (2010). https://doi.org/10.1007/978-0-387-09751-0
3. Schmidt, C.S.: Flexible information discovery with guarantees in decentralized distributed systems. Ph.D. dissertation. University of New Jersey, USA (2005)
4. Huang, X., Chang, Y., Chen, S.: PeerCluster: a cluster-based peer-to-peer system. IEEE Trans. Parallel Distrib. Syst. **17**(10), 1110–1123 (2006)
5. Huang, Y., Du, H., Zhang, G.: Clustering model of P2P CDN based on the prediction of user requirements. JNW **7**(3), 532–539 (2012). Academy Publisher
6. Huang, C., Li, X., Wu, J.: A semantic searching scheme in heterogeneous unstructured P2P networks. Comput. Sci. Technol. **6**(266), 925–994 (2011)
7. Ma, Y., Tan, Z., Chang, G., Gao, X.: A P2P overlay network routing algorithm based on group-average agglomerative clustering topology. In: 9th International Conference on Hybrid Intelligent Systems, vol. 2, pp. 445–448 (2009)
8. Kacimi, M., Y'etongnon, K.: Density-based clustering for similarity search in a P2P network. In: 6th International Symposium on Cluster Computing and the Grid, pp. 57–64 (2006)
9. Datta, S., Giannella, R., Kargupta, H.: Approximate distributed k-means clustering over a peer-to-peer network. IEEE Trans. Knowl. Data Eng. **21**(10), 1372–1388 (2009)
10. Dumitrescu, M., Andoni, R.: Clustering SPs in P2P networks by growing neural gas. In: 18th Euromicro Conference on Parallel, Distributed and Network-Based Processing, pp. 311–318 (2012)
11. Atul, S., Mads, H.: Decentralized clustering in pure P2P overlay networks using Schelling's model. In: IEEE Communications Society Subject Matter Experts for Publication in the ICC (2007)
12. Lakshmish, R., Bugra, G., Ling, L.: Connectivity based node clustering in decentralized peer-to-peer networks. In: Proceedings of the 3rd International Conference on Peer-to-Peer Computing, P2P 2003 (2003)
13. Saurabh, T., Leonard, K.: Optimal search performance in unstructured peer-to-peer networks with clustered demands. IEEE J. Sel. Areas Commun. **25**(1), 84–95 (2007)
14. Ayyasamy, S., Sivanandam, S.N.: A cluster based replication architecture for load balancing in peer-to-peer content distribution. arXiv:1009.4563 (2010)
15. Michalis, V., Kjetil, N., Christos, D.: Peer-to-Peer Clustering for Semantic Overlay Network Generation. INSTICC Press, Setubal (2006)
16. Tirado, M., Higuero, D., Isaila, F., Carretero, J., Iamnitchi, A.: Affinity P2P: a self-organizing content-based locality-aware collaborative peer-to-peer network. Comput. Netw. **54**(12), 2056–2070 (2010)
17. Vakili, G., Khorsandi, S.: Self-organized cooperation policy setting in P2P systems based on reinforcement learning. IEEE Syst. J. **7**(1), 151–161 (2013)
18. Ebrahimi, M., Rouhani, R., Seyed, M.: An ant-based approach to cluster peers in P2P database systems. Knowl. Inf. Syst. **43**(1), 219–247 (2015)
19. Meng, X.: A churn-aware durable data storage scheme in hybrid P2P networks. J. Supercomput. **74**(1), 183–204 (2018)
20. Xianfu, M., Jing, J.: A free rider aware topological construction strategy for search in unstructured P2P networks. Peer-to-Peer Netw. Appl. **9**(1), 127–141 (2016)

21. Zungeru, M., Ang, M., Seng, P.: Termite-hill: from natural to artificial termites in sensor networks. Int. J. Swarm Intell. Res. **3**(4), 1–23 (2012)
22. Martin, H.R.: Termite: a swarm intelligent routing algorithm for mobile wireless ad-hoc networks. Ph.D. dissertation, Electrical and Computer Engineering, Cornell University, NY, United States (2005)
23. Marco, D., Thomas, S.: Ant Colony Optimization. MIT Press, Cambridge (2004)
24. Selcuk, O., Dervis, K.: Routing in wireless sensor networks using an ant colony optimization (ACO) router chip. Sensors **9**, 909–921 (2009)
25. PeerSim. http://peersim.sourceforge.net/

Performance and Energy Consumption Prediction of Randomly Selected Nodes in Heterogeneous Cluster

Sara Kadhum Idrees⬛ and Ahmed B. M. Fanfakh$^{(\boxtimes)}$⬛

Computer Department, College of Sciences for Woman, University of Babylon,
Hillah, Iraq
sarakidrees@gmail.com, afanfakh@gmail.com

Abstract. The heterogeneous computing cluster consists of a number of nodes that are different in their computing powers and their consumed energy. Executing parallel application over heterogeneous cluster produces various tasks' execution times according to the heterogeneity of the nodes' computing powers. Consequently, different task execution times in a synchronous application are wasting the energy consumed of the underlying parallel platform running them. However, the execution time prediction process of parallel applications over a dynamically selected set of heterogeneous nodes is significant in the area of parallel computing. The predicted execution time almost uses in the process of optimizing both the speedup and the consumed energy of the parallel application running on the selected set of heterogeneous nodes in the cluster. Therefore, prediction helps the algorithm to select in advance the best solution without the need for trial and error methods or other costly methods. This paper presents new models to predict the execution time and consumed energy of parallel application when a random set of heterogeneous nodes are selected. The performance and energy consumption results are predicted for message passing synchronous applications. Moreover, the experimental results are evaluated over SimGrid/SMPI simulator. The results show that the proposed prediction models give a very acceptable precision when they are compared with the real measurements.

Keywords: Performance prediction · Energy consumption prediction
MPI applications · Heterogeneous cluster

1 Introduction

Heterogeneous cluster consists of a group of nodes that are different from where the computing power and the energy consumption. The heterogeneity in the computing power of nodes lead to imbalanced workload when executing the parallel message passing programs over that heterogeneous platform. Imbalanced workloads produce idle times that happen when fast nodes suspense the slowest nodes. However, idle times increase both executing time and the consumed energy of the parallel application.

Thus, the goal of many programming techniques and models is to minimize the execution time of the parallel programs and the consumed energy of the platform

© Springer Nature Switzerland AG 2018
S. O. Al-mamory et al. (Eds.): NTICT 2018, CCIS 938, pp. 21–34, 2018.
https://doi.org/10.1007/978-3-030-01653-1_2

executing them. Accordingly, the metrics used to evaluate program performance for parallel or distributed programs, such as speedup and efficiency, are mainly based on the execution time of the program. Nowadays, program efficiency has been expanded to include energy consumption into consideration [1]. Indeed, this is very important for large and complex software systems that are used in many fields of research and industry. However, many optimization techniques in the literature depend on the execution time prediction methods that help researchers in the process of making decisions for each new state in the system. The prediction methods can be implemented using many tools such as: statistic tools, AI algorithms, heuristic methods and analytical mathematical modeling. Some of these tools are costly in terms of time complexity when many iterations are needed to predict the running time of the parallel application.

In this paper, a new analytical prediction model for performance and energy consumption running over heterogeneous platform are proposed. They are dynamically predicting the execution time and the energy consumption for any number of heterogeneous nodes. The accuracy of these models was evaluated by running some message passing programs over heterogeneous clusters, each with different nodes in term of their hardware types and numbers.

The remainder of the paper is structured as follows: Sect. 2 describes a review of some related works. Section 3 delineates performance and energy model for heterogeneous platform. Section 4 describes the proposed prediction methods for performance and energy consumption. Section 5 shows the experimental results. The paper terminates with the conclusion in Sect. 6.

2   Related Works

Authors focus on some tools for predicting the execution time of parallel applications such as statistic methods, heuristic algorithms and mathematical models. Recently, energy consumption is an important factor considered in the area of green computing. It is affected directly by the execution time of the parallel application. If the parallel application's execution time is increased, then propositionally the consumed energy by the parallel hardware running the application is increased too. However, this section presents some related works that are interested in predicting the execution time of the message passing parallel application, in addition to the works that predict the energy consumption of these applications.

There are several statistical and analytical models that have been proposed to predict the performance of HPL (High Performance Linpack) benchmark to achieve maximum possible performance. To predict the running time of HPL benchmark, researchers proposed an analytical model in [2]. To predict maximum possible performance, they propose a semi-empirical model. An optimization for the existing model is proposed by partitioning the performance modeling into message passing overhead, computational cost and proposed communication method as in [3]. The prediction error was less than 5% after implementing that model on different cluster platforms. In [4], researchers found the parameters of the HPL benchmark that can be used to create power consumption and performance prediction model. To predict the performance of the parallel application, authors illustrate piecewise polynomial regression and (ANNs) in [5]. In [6], researchers presented a machine learning approach using multilayer neural networks. In [7] researchers explored novel regression-based approaches to predict parallel program scalability. These techniques provide a good prediction accuracy between 6.2% and 17.3%. Authors in [8], introduced adaptive model for predicting the execution time, i.e. Time-to-Solution (TtS) of parallel applications that consider the problem size of the application but with fixed number of compute nodes. Even though, by using the corresponding knowledge of (TtS), it could be possible to extract the energy consumption of applications. In [9], a hybrid method for execution time prediction has been presented. It combines historic based prediction techniques with profile-based ones and essentially aims to provide scheduling algorithms with information about the submitted applications to schedule them. The hybrid method for execution time prediction has been evaluated by the MPI parallel application. To predict the performance of large-scale applications, authors in [10] developed a prediction framework that run the parallel applications using a single small-scale training. In [11], researchers illustrate accurate prediction method for large-scale parallel applications. They proposed a novel approach to predict the sequential computation time accurately and efficiently. The proposed approach only needs an individual node of a target platform but the overall target platform does not need to be available. Many of the analytical models for performance prediction of parallel applications built on the target platform in [12–16]. The primary advantage of analytical approaches is the low cost of its implementation. However, analytical model construction requires an understanding of the algorithms and their practical implementations. For scientific applications, authors proposed the performance predictions model that implemented on a homogeneous cluster of workstations [17]. They developed automatic parallel application prediction system for cluster computing environments. Authors in [18], proposed a method for predicting the performance and power consumption under various control knobs combinations of a parallel application. They were capable to predict its behaviour in all the other settings by using linear regression model and exploring a few [Frequency and Threads] configurations. In [19], authors proposed a system that tackles different configurations at runtime to find the best one in terms of performances and power consumption that satisfies the given requirements. Other works [20] and [21], presented more general models allow the prediction after implementing different configurations of different applications. In [22], by using only one node of a homogeneous cluster, authors proposed models that can be used to predict in the energy consumption of MPI applications and used the proposed computation and communication time models as a basis in prediction. Researchers in [23] presented prediction model for the consumed energy when collective MPI application is used for large scale HPC systems. To predict the power and energy consumption of parallel HPC applications for different number of compute nodes, authors proposed Adaptive Energy and Power Consumption Prediction (AEPCP) model [24]. In [25], authors proposed an analytical model that simulate the energy consumption effect of frequency scaling. In [26], authors applied the model to parallel tasks that are executed concurrently with each other by a set of processors.

In addition to the previous works, this paper uses analytical models to predict the execution time and the consumed energy of parallel message passing applications executed on heterogeneous cluster. The process of using analytical models is economic from their very small execution time to give the results. Especially, when compared to the other methods such as AI methods that need training or regression methods that depend on executing many iterations.

3 The Performance and Energy Model for Heterogeneous Cluster Platform

3.1 The Execution Time Model for Heterogeneous Cluster

Message passing interface (MPI) parallel programs implemented over wide types of parallel platforms. It has the ability of portably executing its communication routines over the underline parallel architecture. Generally, MPI programs consist of two parts: computations and communications parts. The former refers to the amount of arithmetic operations, comparisons and others that do not belong to any communications. The latter involving send and receive operations that coordinate parallel tasks communications. Moreover, heterogeneous cluster is a group of non-equivalent computing nodes where each one is different in the computing power and the consumed energy from other nodes. All nodes are interconnected together via homogeneous networks, all links have similar bandwidth and latency. Therefore, the execution of parallel task over heterogeneous cluster result in various computation times even if the problem is load balance when it decomposed. Before all nodes are capable to communicate with each other in a synchronous manner, slack times may happen if the fast nodes have to wait for the slowest node to finish its computations, see Fig. 1. However, the execution time of parallel program is the execution time of the slower task.

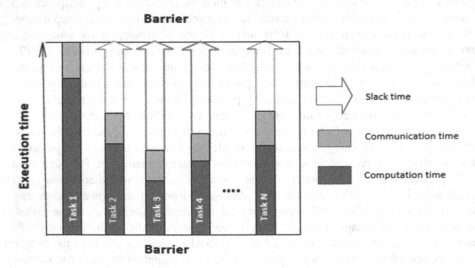

Fig. 1. Parallel tasks execution over heterogeneous cluster

Hence, the running time of message passing parallel applications is computed according to the slowest task that has the maximum computation time plus the minimum communication time, which does not have slack time as in Eq. 1, for more detail please refer to [27].

$$T_{parallel} = \max_{i=1,2,...,N} (T_{cp_i}) + \min_{i=1,2,...,N} (Tcm_i) \quad (1)$$

Where T_{cp_i} and Tcm_i are the computation and the communication times of the node i respectively. This model captures the running time of the parallel program by computing the maximum computation time added to the time of communication of the slowest node without any slack time.

3.2 The Energy Consumption Model of Heterogeneous Cluster

The consumed power of a processor can be partitioned into two power metrics: dynamic and static power [28–32]. The first one is consumed during the computation times and the latter is consumed throughout all times when the processor is turned on [29]. In the heterogeneous cluster, each node has different dynamic and static powers, denoted as P_{dyn} and P_{static} respectively. Therefore, each node may have different energy consumption, according to its power consumption. While, the energy consumption is the product of the power consumption of the node multiplied by the execution time of that node. Consequently, each node will consume different energy due to its execution times that are variable from each other.

The energy consumed of the individual processor in a heterogonous cluster can be computed according to its characteristics (its time of computations and both dynamic and static powers). Then, processor individual energy consumption computed as follows:

$$E_{processor} = (P_{dyn} + P_{static}).T \quad (2)$$

Where T is the execution time for all parallel program.

In the case of parallel systems, all nodes have to communicate with each other during the synchronous execution. Thus, the node's processor will remain idle during the communications time and consumed only static power.

However, the total energy consumed by a heterogeneous cluster composed of N nodes equal to the total energy consumed by each node.

Formally, according to [27], the energy consumed by heterogeneous cluster $E_{cluster}$ including all nodes' processors executing the message passing distributed applications can be expressed as follows:

$$E_{cluster} = \sum_{i=1}^{n} E_{processor_i} = \sum_{i=1}^{n} P_{dyn_i}.Tcp_i + \sum_{i=1}^{n} P_{static_i}.T_{parallel} \quad (3)$$

Where Tcp_i is the computation time of node i and $T_{parallel}$ is the execution time of parallel application.

The next section presents the proposed prediction models for performance and energy consumption that apply to any parallel message passing application executed over heterogonous cluster.

4 The Proposed Prediction Methods for Performance and Energy Consumption on Heterogeneous Architecture

The execution time of parallel applications is affected directly by the computing power of the parallel machine executing them. Therefore, they are linearly propositional if the application is composed of independent tasks where there is no dependency and no communications between them. This situation does not always exist in the problems solved in the parallel computing. Moreover, communications are the main bottleneck in the distributed parallel programs. They increase drastically the execution time of the application when they are increased. Recently, the performance of parallel application becomes not only the unique factor required in the area of scientific parallel applications. The wide increase in the number of computing units of parallel architecture has increased the energy consumed by the platform implicating them. Thus, the performance plus the energy consumption factors must be considered together in the design of parallel clusters and the applications executed over them in parallel. In the literature, authors explain that the execution time is effectively increased the consumed energy when it increased, see [25].

Therefore, the process of predicting both the execution time and energy consumption is a very important process, which aids in the designing efficient parallel applications in term of energy consumption and performance. However, the next subsections describe the proposed method of predicting both the running time and the consumed energy of message passing parallel applications.

4.1 The Execution Time Prediction Method

The main purpose of the prediction method is modeling the behaviour of a dynamic environment. Especially, if one or more factors affecting it are changed. In this work, we are interested in executing parallel message passing applications over heterogeneous cluster. As mentioned previously, a heterogonous cluster composed from N nodes. The latter's processors have different types in terms of computing power and power consumption. Moreover, any change in the number of nodes or processors types may directly affect the execution time of the parallel application running on them. Therefore, the goal of this section is to develop an analytical model which predicts the execution time of parallel application that executes over any number of selected nodes in the cluster, which may have different types. Every node has different computing power (FLOPS), and then each one has different computation time. The prediction method depends on a number of steps to predict the execution time. These steps are demonstrated by a set of analytical equations as in the following:

We assume if the parallel application executes over any number of heterogeneous nodes, denoted as NRS, then different computation times, Tcp_i will be produced according to the differences in the nodes' computing powers. However, the execution time of the parallel application is changed. The execution time of message passing parallel application consists of computation and communication times. Thus, the first step in our method is gathering these times by running the application once over all types of nodes, one node per type. The number of nodes types in the heterogeneous cluster is denoted as NT. Depending on the computation time values gathered from nodes, one per type, the serial computation time, and the time of executing the application over the slowest node, is computed by multiplying the computation time of the slower task by number of nodes types NT as follows:

$$serial_{tcp} = T_{cpslower} \cdot NT \qquad (4)$$

Where $T_{cpslower}$ is the computation time for the slowest node.

Suppose the application executed over a homogenous cluster with NRS nodes. However, the homogenous computation time over each node can be computed by dividing the serial computation time, Eq. (4), of the parallel application by the number of the node NRS of the cluster as follows:

$$T_{cp_{homo}} = \frac{serial_{tcp}}{NRS} = \frac{T_{cpslower} \cdot NT}{NRS} \qquad (5)$$

Where NRS is the number of random selected nodes.

Consequently, to find the percentage of change between computations times of an application running over a heterogeneous cluster composed of NT nodes. The heterogeneous computing factor HCF is computed as follows:

$$HCF_i = \frac{T_{cpslower}}{T_{cp_i}} = \frac{\max\limits_{i=1,2,...,NT}(T_{cp_i})}{T_{cp_i}} \qquad (6)$$

Where HCF_i is the heterogeneous computation factor of node i.

Depending on these factors which computed for each node type, the predicted heterogeneous computation time for each node in the cluster composed of NRS nodes is computed. This is by multiplying the homogenous computation time by the heterogeneous computing factor HCF for the node j in a cluster composed from NRS nodes of type t, as follows:

$$T_{cp_{predicted_j}} = T_{cp_{homo}} \cdot HCF_{tj} = \frac{T_{cpslower} \cdot NT}{NRS} \cdot HCF_{tj} \qquad (7)$$

However, the predicted computation times are increased or decreased by a factor of HCF.

The running time of synchronous parallel application is the running time of the slowest task. It consists of computation and communication sections. Thus, to predict the execution time of the message passing program both computation time and communication times must be predicted. Therefore, the computation time of the program is the computation time $T_{cps_{predicted}}$ over the slower node, that is computed as follows:

$$T_{cps_{predicted}} = \max_{j=1,2,\ldots,NRS} T_{cp_{predicted\,j}} \tag{8}$$

Then, the communication time of the slower task is proportionally increased or decreased if the number of nodes is increased or decreased too. The communication of the slower task, T_{cm}, from the execution of program over cluster of NT nodes proportionally computed for the new cluster with NRS nodes as in Eq. (9). The predicted communication time of the slowest task is the communication time without slack times as below:

$$T_{cm_{predicted}} = \frac{T_{cm} \cdot NRS}{NT} \tag{9}$$

Finally, depending on Eqs. (8) and (9), the predicted execution time for the parallel synchronous distributed application computes the execution time of the slowest node as in Eq. (10).

$$T_{predicted} = \max_{j=1,2,\ldots,NRS} T_{cp_{predicted\,j}} + \frac{T_{cm} \cdot NRS}{NT} \tag{10}$$

The proposed model (10), is used to predicate the execution time of the any parallel message passing application over any set of heterogeneous nodes. It only depends on the gathered information from the execution of the application once over a set of nodes, which is one node per type in heterogynous cluster.

4.2   The Energy Consumption Prediction Method

In the heterogeneous platform, each node has different dynamic and static powers depending on some parameters such as processor frequency, number of cache levels, main memory frequency etc. Therefore, the energy consumption of the synchronous distributed MPI application is different from each other. While the energy consumption is proportional with execution time of an application. However, the precision of the proposed energy prediction model mainly depends on the computation time model as in Eq. (7) and the execution time of the parallel program as in Eq. (10). Indeed, the static energy is consumed overall the execution time and the dynamic energy is consumed during time of computation only. Hence, the work presented in this paper is mainly dependent on the time model. Thus, the prediction energy model for a message passing parallel application executed over any selected number of heterogeneous nodes can be formulated as follows:

$$E_{total\,predicted} = \sum_{i=1}^{NRS} P_{dyn_i} \cdot T_{cp_{predicted_i}} + \sum_{i=1}^{NRS} P_{ststic_i} \cdot T_{predicted} \tag{11}$$

Where $T_{cp_{predicted_i}}$ is computed as in the Eq. (7) and $T_{predicted}$ is computed as in the Eq. (10).

5 Experimental Results

The proposed method of this work is to predict the execution time and the energy consumption of message passing programs executed over any set of heterogeneous nodes. Therefore, the next section describes the experimental configuration and results.

5.1 Experiments Configuration

To test the ability of the proposed method in the prediction of both the execution time and the energy consumption of message passing program. The heterogeneous cluster executing these applications was simulated using SimGrid/SMPI simulator [33] to apply our experiments on it. The SimGrid/SMPI simulator has flexible tools to create a cluster architecture and executes message passing applications on it. The simulator was configured to use a heterogeneous cluster with four different types of nodes. Each node in the cluster has different characteristics from the others such as computing power (FLOPS), the dynamic and static powers. Nodes are connected through an Ethernet network of 10 Gbit/s bandwidth. Table 1 shows the detailed characteristics of these four nodes.

Table 1. Characteristics of heterogeneous nodes.

Node type	Simulated GFLOPS	Processor freq. (GHz)	Dynamic power (W)	Static power (W)
1	40	2.50	20	4
2	50	2.66	25	5
3	60	2.90	30	6
4	70	3.40	35	7

5.2 Simulation Results

This section verifies the accuracy of the proposed execution time and the energy prediction models Eqs. (10) and (11) respectively. To be able to evaluate the results of these models, three parallel message-passing interface (MPI) applications were built. These parallel applications are 2D matrix multiplication, iterative Jacobi method and merge sort. In this paper, we are denoted to these applications as MM, Jacobi and Msort respectively. The size of MM application is $10^3 \times 10^3$, Jacobi is $10^3 \times 10^3$ and Msort is 10^8. These applications are balanced and have different computations and communications ratios. Five scenarios proposed to represent different number of nodes which are 8, 16, 32, 64, 128 nodes scenarios. The proposed method starts generating a number of heterogeneous nodes randomly each with different types. For example, if the 32 nodes scenario is used, then an algorithm starts to select random heterogeneous nodes until reaching to 32 nodes. The parallel message passing program executed twice. The first one is by executing the application over 4 nodes, one node per type, where 4 types are existing in this experiment. In this case, the computation and the communication times gathered to predict the execution time and the consumed energy using Eqs. (7) and (10) respectively. After the predicted execution time and the energy consumption was computed, the second execution is started with a cluster containing exactly the same number of nodes and types. The real execution time for the parallel program and its energy is computed as in Eq. (3). Both real and predicted execution times are compared in Fig. 2. These figures present results of three programs: MM, Jacobi and Msort respectively. The results in the figures are the average of multiple runs for each scenario.

Figures show that the proposed execution time prediction model gives good accuracy results for all three programs. On average, the percentage errors are 3.66, 2.82 and 0.38 for programs MM, Jacobi, Msort respectively. Figure 3 presents the results of the real and the predicted energy consumption. They were computed using Eqs. (3) and (7) that depends on the dynamic and the static power values presented in Table (1). They show also good accuracy for all programs. The average percentage errors were 7.31, 0.38 and 13.84 for programs MM, Jacobi and Msort respectively. According to these results, the used message passing programs have different computation to communication ratios. These differences produced variable error ratios between the real and predicted values for each program as shown in the figures.

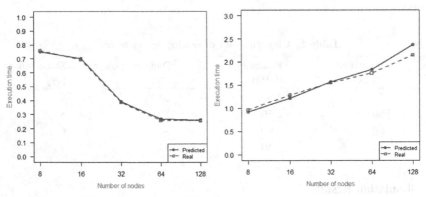

(a)The MM program(b) The Jacobi program

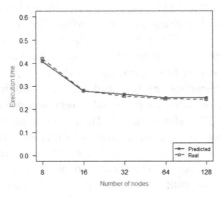

(c)The Msort program

Fig. 2. The execution time results

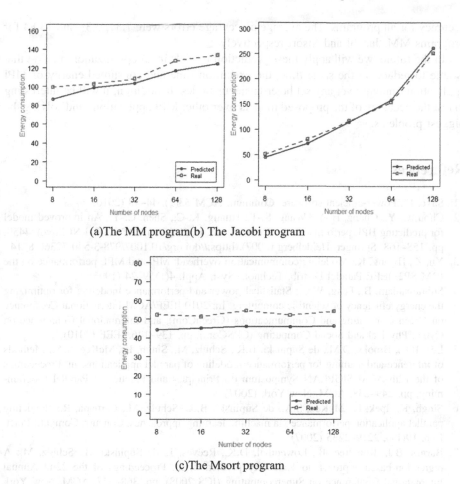

(a)The MM program(b) The Jacobi program

(c)The Msort program

Fig. 3. The energy consumption results

6 Conclusion and Future Works

The process of predicting the running time and the consumed energy of parallel application over a set of heterogeneous nodes has important challenge. However, this paper presents new prediction models for running time and energy consumption of the MPI programs running on any set of heterogeneous nodes. To evaluate the proposed methods on a heterogeneous cluster platform, the models applied on the matrix multiplication, were iterative Jacobi method and merge sort application. These applications are load balanced and have different computations and communications ratios. They executed over the SimGrid/SMPI simulator. Experiments showed that the proposed execution time prediction model gives good accuracy results for all three programs. On average, the percentage errors are 3.66, 2.82 and 0.38 for programs MM, Jacobi, Msort respectively. Also, the prediction model for energy consumption showed a good accuracy for all programs. The average percentage errors were 7.31, 0.38 and 13.84 for programs MM, Jacobi and Msort respectively.

In the future, we will apply these prediction models in an optimization method that is able to reduce, at the same time, the execution time and consumed energy of MPI application running over any set heterogeneous nodes. In addition, it will be interesting to test the accuracy of the proposed models over other MPI applications and solving the biggest problems.

References

1. Saxe, E.: Power-efficient software. Commun. ACM **53**(2), 44–48 (2010)
2. Chou, C.-Y., Chang, H.-Y., Wang, S.-T., Huang, K.-C., Shen, C.-Y.: An improved model for predicting HPL performance. In: Cérin, C., Li, K.-C. (eds.) GPC 2007. LNCS, vol. 4459, pp. 158–168. Springer, Heidelberg (2007). https://doi.org/10.1007/978-3-540-72360-8_14
3. Xu, Z., Hwang, K.: Modeling communication overhead: MPI and MPL performance on the IBM SP2. IEEE Parallel Distrib. Technol., Syst. Appl. **4**(1), 9–24 (1996)
4. Subramaniam, B., Feng, W.C.: Statistical power and performance modeling for optimizing the energy efficiency of scientific computing. In: 2010 IEEE/ACM International Conference on Green Computing and Communications (GreenCom), and International Conference on Cyber, Physical and Social Computing (CPSCom), pp. 139–146. IEEE (2010)
5. Lee, B.C., Brooks, D.M., de Supinski, B.R., Schulz, M., Singh, K., McKee, S.A.: Methods of inference and learning for performance modeling of parallel applications. In: Proceedings of the 12th ACM SIGPLAN Symposium on Principles and Practice of Parallel Programming, pp. 249–258. ACM, New York (2007)
6. Singh, K., Ipek, E., McKee, S.A., de Supinski, B.R., Schulz, M., Caruana, R.: Predicting parallel application performance via machine learning approaches. Concurr. Comput.: Pract. Exp. **19**(17), 2219–2235 (2007)
7. Barnes, B.J., Rountree, B., Lowenthal, D.K., Reeves, J., de Supinski, B., Schulz, M.: A regression-based approach to scalability prediction. In: Proceedings of the 22nd Annual International Conference on Supercomputing (ICS 2008), pp. 368–377. ACM, New York (2008)
8. Hennessy, J.L., Patterson, D.A.: Computer Architecture: A Quantitative Approach. Elsevier, New York (2012)
9. Miegemolle, B., Monteil, T.: Hybrid method to predict execution time of parallel applications. In: Proceedings of the 2008 International Conference on Scientific Computing, CSC 2008, Las Vegas, Nevada, USA, pp. 224–230 (2008)
10. Jayakumar, A., Murali, P., Vadhiyar, S.: Matching application signatures for performance predictions using a single execution. In: 2015 IEEE International Parallel and Distributed Processing Symposium (IPDPS), pp. 1161–1170. IEEE (2015)
11. Zhai, J., Chen, W., Zheng, W.: PHANTOM: predicting performance of parallel applications on large-scale parallel machines using a single node. In: ACM SIGPLAN Notices, pp. 305–314. ACM (2010)
12. Berker, K.J, Pakin, S., Kerbyson, D.J.: A performance model of the Krak hydrodynamics application. In: International Conference on Parallel Processing (ICPP 2006), Columbus, Ohio, pp. 245–254 (2006)
13. Hoisie, A., Lubeck, O., Wasserman, H.: Performance and scalability analysis of teraflop-scale parallel architectures using multidimensional wavefront applications. Int. J. High Perform. Comput. Appl. **14**(4), 330–346 (2000)

14. Kerbyson, D.J., Alme, H.J., Hoisie, A., Petrini, F., Wasserman, H.J., Gittings, M.: Predictive performance and scalability modeling of a large-scale application. In: Proceedings of the 2001 ACM/IEEE conference on Supercomputing, Denver, CO, pp. 37–48 (2001)

15. Mathis, M.M., Kerbyson, D.J., Hoisie, A.: A performance model of non-deterministic particle transport on large-scale systems. In: Sloot, P.M.A., Abramson, D., Bogdanov, A.V., Gorbachev, Y.E., Dongarra, J.J., Zomaya, A.Y. (eds.) ICCS 2003. LNCS, vol. 2659, pp. 905–915. Springer, Heidelberg (2003). https://doi.org/10.1007/3-540-44863-2_89

16. Charr, J.C., Couturier, R., Fanfakh, A., Giersch, A.: Dynamic frequency scaling for energy consumption reduction in synchronous distributed applications. In: ISPA 2014: The 12th IEEE International Symposium on Parallel and Distributed Processing with Applications, pp. 225–230. IEEE Computer Society, Milan (2014)

17. Khan, R.Z., Ansari, A.Q., Qureshi, K.: Performance prediction for parallel scientific application. Malays. J. Comput. Sci. 17(1), 65–73 (2004)

18. De Sensi, D.: Predicting performance and power consumption of parallel applications. In: 2016 24th Euromicro International Conference on Parallel, Distributed, and Network-Based Processing (PDP), pp. 200–207. IEEE (2016)

19. Li, J., Martinez, J.F.: Dynamic power-performance adaptation of parallel computation on chip multiprocessors. In: Proceedings of 12th International Symposium on High-Performance Computer Architecture, pp. 77–87. IEEE, Austin (2006)

20. Curtis-Maury, M., Shah, A., Blagojevic, A., Nikolopoulos, D.S., de Supinski, B.R., Schulz, M.: Prediction models for multi-dimensional power-performance optimization on many cores. In: Proceedings of the 17th International Conference on Parallel Architectures and Compilation Techniques, pp. 250–259. ACM (2008)

21. Cochran, R., Hankendi, C., Coskun, A., Reda, S.: Identifying the optimal energy-efficient operating points of parallel workloads. In: Proceedings of the International Conference on Computer-Aided Design, pp. 608–615. IEEE Press (2011)

22. Heinrich, F., et al.: Predicting the Energy Consumption of MPI Applications at Scale Using a Single Node (2017)

23. Diouri, M.E.M., Glück. O., Mignot. J.C., Lefèvre. L.: Energy estimation for MPI broadcasting algorithms in large scale HPC systems. In: Proceedings of the 20th European MPI Users' Group Meeting, pp. 111–116. ACM (2013)

24. Shoukourian, H., Wilde, T., Auweter, A., Bode, A.: Predicting the energy and power consumption of strong and weak scaling HPC applications. Supercomput. Front. Innov. 1(2), 20–41 (2014)

25. Rauber, T., Rünger, G.: Analytical modeling and simulation of the energy consumption of independent tasks. In: Proceedings of the Winter Simulation Conference, p. 245 (2012)

26. Rauber, T., Rünger, G.: Modeling the energy consumption for concurrent executions of parallel tasks. In: Proceedings of the 14th Communications and Networking Symposium, pp. 11–18. Society for Computer Simulation International (2011)

27. Charr, J.C., Couturier, R., Fanfakh, A., Giersch, A.: Energy consumption reduction with DVFS for message passing iterative applications on heterogeneous architectures. In: 2015 IEEE International Parallel and Distributed Processing Symposium Workshop (IPDPSW), pp. 922–931. IEEE Computer Society, India (2015)

28. Le Sueur, E., Heiser, G.: Dynamic voltage and frequency scaling: the laws of diminishing returns. In: Proceedings of the 2010 International Conference on Power Aware Computing and Systems, pp. 1–8 (2010)

29. Malkowski, K.: Co-adapting scientific applications and architectures toward energy-efficient high-performance computing. The Pennsylvania State University, USA (2008).

30. Zhuo, J., Chakrabarti, C.: Energy-efficient dynamic task scheduling algorithms for DVS systems. ACM Trans. Embed. Comput. Syst. 7(2), 1–25 (2008)

31. Rizvandi, N.B., Taheri, J., Zomaya, A.Y.: Some observations on optimal frequency selection in DVFS-based energy consumption minimization. J. Parallel Distrib. Comput. **71**(8), 1154–1164 (2011)
32. Zhuo, J., Chakrabarti, C.: Energy-efficient dynamic task scheduling algorithms for DVS systems. ACM Trans. Embed. Comput. Syst. (TECS) **7**(2), 17 (2008)
33. SimGrid: Versatile Simulation of Distributed Systems. http://www.simgrid.gforge.inria.fr

Data Transmission Protocol for Reducing the Energy Consumption in Wireless Sensor Networks

Rafal Alhussaini[iD], Ali Kadhum Idrees[✉][iD],
and Mahdi Abed Salman[iD]

Department of Computer Science, University of Babylon, Babylon, Iraq
rafalalhussaini@yahoo.com,
{ali.idrees,mahdi.salman}@uobabylon.edu.iq

Abstract. One of the crucial factors in designing Wireless Sensor Networks (WSNs) is energy management. In order to save the limited energy of the WSNs, it is essential to reduce the redundant data during the aggregation inside the sensor node. This can lead to eliminating the communication cost related to sending unnecessary data. Therefore, it is necessary to implement a data reduction method inside each sensor node to remove non-useful data before transmitting it to the sink. This paper proposed Data Transmission (DaT) Protocol for Reducing the data transmission cost inside each sensor node by getting rid of the redundant data to save the energy while maintaining a suitable level of accuracy in the received readings at the sink. DaT protocol partitions the WSN lifetime into periods. Two stages are found in every period. The first stage is the data classification in which the collected sensed data are classified into several classes based on the Modified k-Nearest Neighbor technique. The similar classes are merged into one class to further reduce the data size transmitted to the sink. In the second stage, the best representative reading from every class is selected to send it to the sink. DaT protocol is evaluated using real recorded data of the temperature and using OMNeT++ network simulator to show the performance of the proposed protocol. Compared with other recent methods, DaT protocol can reduce the energy consumption, the data loss ratio, and the size of the data after applying data reduction.

Keywords: Wireless Sensor Networks (WSNs) · Periodic applications
Data reduction · Modified k-Nearest Neighbour · Energy efficiency

1 Introduction

Every day many smart devices are connected to the international network, where the IoTs era is coming strongly. The IoT refers to a large number of physical devices (like sensors, home appliances, actuators, vehicles, etc.) which are connected by a network to exchange the data resulting in performance enhancements, economic advantages, and decreased human efforts [17]. The IoTs networks will impose a big challenge for

© Springer Nature Switzerland AG 2018
S. O. Al-mamory et al. (Eds.): NTICT 2018, CCIS 938, pp. 35–49, 2018.
https://doi.org/10.1007/978-3-030-01653-1_3

decision-makers due to the huge data produced by these large networks. One of the big data sources in the IoTs network is the Wireless sensor network (WSNs) [17, 26, 30]. WSNs became one of the most active research areas in recent years due to the wide range of applications that utilize this type of network such as environment and metrology monitoring to industrial, security, health and social applications [1, 24]. WSNs are a special class of network consisting of a large number of small size, low-power and low-cost smart devices called sensor nodes. A large number of nodes are usually deployed over a wide geographical area [25, 27]. Each node collects data from its surrounding environment [2]. The principal challenge in the WSNs is how to manage the huge collected data in an energy-efficient way [20, 31].

For example, if the system monitors some meteorological conditions such as temperature or humidity. The registered record of these parameters has to be transmitted at certain intervals. However, depending on the application some and, usually much, of these data are redundant. Therefore, the amount of data can be reduced without affecting the quality of required information or at least the amount of lost data is within the required accuracy of the system. Rather than sending every registered parameter as it is recorded, a group of them is either stored locally in the sensor node then released at certain intervals or as per request from the end node, or a number of these measurements are collected from a number of neighbouring nodes by a head node and passed over to the end node (sink node).

This paper gives the following contributions:

(1) Data Transmission (DaT) Protocol for Reducing the Energy Consumption in WSNs through using the Modified k-Nearest Neighbour is proposed. DaT reduces data transmission inside each sensor node by selecting the best representative data in each period instead of sending all the data to the sink. Rapid grouping technique is used to select initial sensed data (seeds) at random which are then compared to the rest of the sensed data in the period to assign each sensed data to the closest seed. Once a predefined number of initial classes are formed, close clusters are then combined to further reduce the data size.

(2) DaT protocol is evaluated using real recorded data of the temperature and using OMNeT++ network simulator to show the performance of proposed protocol.

(3) The results are compared with other recent methods like ATP (Harb et al. [18]) and the Prefix-Frequency Filtering (PFF) technique (Bahi et al. [19]). DaT protocol can reduce the communication cost and save energy and thus improve the network lifetime.

The rest of the paper is organized as follows. Section 2 introduces different data aggregation approaches as related works. Section 3 presents the proposed DaT protocol. Simulation results are illustrated in Sect. 4. Section 5 concludes and future works directions.

2 Related Works

Different strategies have been explored in order to reduce the amount of energy spent on data transmission by reducing the amount of data to be sent from the sensor node to the sink node. Data aggregation is one popular scheme in reducing transmitted data [3–7]. Data compression is also one of the other tools that are being explored as a means to minimize energy consumed by the data transmission [8, 9, 11, 15]. Utilizing data prediction techniques is yet another popular strategy to remove redundant data thus reducing the amount of data transmission [3, 7, 8, 14, 28, 29]. The common goal in all these approaches is to reduce the volume of transmitted data without significantly affecting the amount of information required by the receiving sink.

Linear prediction coding (LPC) has been adopted for data compression in [8]. The LPC is used to reduce the data by exploiting the spatial correlation in the data. Orthogonal Matching Pursuit (OMP) is then deployed to recover the original data. The authors in [9] utilized the lightweight temporal compression algorithm (LCT) as a tool for data reduction between sensor nodes and sink node. The algorithm was applied to data collected from mica motes. It was shown that using this algorithm they could achieve up to 20% compression ratio. The work in [11] summarizes different data compression schemes and concludes that no one scheme is universally superior to others. It is mostly a matter of application and what assumption is being made in each scheme. The hierarchical Least Mean Square (LMS) adaptive filter is employed in [3] as a way of data prediction to reduce power consumption. The algorithm is applied at both the source and the sink and thus the sensor nodes need only to send readings that significantly differ from the predicted value. Predicting future values in a data with high entropy usually associated with the non-linear system is addressed in [7]. Examples of such system are measuring gearbox vibration or electric engine vibration. A model called Self-Exciting Threshold Autoregressive (SETAR) is designed to address the issues related to this kind of application as a method of time series forecasting combined with non-linear systems. It was reported that applying this algorithm on real world data delivered a maximum saving in energy that reaches 73%. The work in [15] presents an effective scheme for data reduction in which a threshold is set to decide if a new reading is to be sent or not based on whether the difference from previous readings is larger or smaller than this threshold. The method was applied to data of heartbeat collected from patients.

A similarity measure based on Kruskal-Wallis Model (KWM) is adopted in [16] as a way of data reduction. The scheme divides reading of certain period into groups that satisfy the KWM measure of similarity between their samples and sends only one reprehensive from each group. This method is applied to the data collected from telosB sensor network. It reported data reduction up to 80% and at the same time maintaining the integrity of transmitted data. The Prefix-Frequency Filtering (PFF) method is introduced in [8] to aggregate the data in WSNs. PFF performs the data aggregation at two levels. The first level applied Jaccard similarity to eliminate the redundant data inside the sensor device. In the second level, the spatial correlation among the sensor nodes is utilized and applied at the aggregator to further remove the data redundancy before sending it to the sink. The authors in [18] proposed an Aggregation and Transmission Protocol (ATP) to minimize the data sending to the sink and save the power at the sensor node. ATP consists of two stages: aggregation and transmission. It detects the correlation among the sensed data to remove the redundant data at the aggregation stage. In the transmission stage, ATP utilized the statistical approaches such as ANOVA model and Fisher test to detect the similarity among the periodically collected data to minimize the sending number of data to the sink.

Shortcomings. Although several papers were proposed for data reduction in WSNs, none of them ensure a high elimination ratio for redundant data while maintaining a suitable level of data accuracy for the received sensed data at the sink. For instance, the compression-based methods can provide a high level of data reduction without ensuring the acceptable level of data accuracy. These methods reduce the consumed power but miss a lot of data that effect on the decision that will be taken at the base station. In the prediction-based methods, the provided data accuracy is suitable without ensuring a high data reduction. This can lead to high energy consumption and thus decreased network lifetime. Therefore, it is necessary to propose a method that balances between the data reduction ratio and the data accuracy in WSNs.

This article suggested Data Transmission (DaT) Protocol saves the power of the sensor nodes by using the Modified k-Nearest Neighbour method to improve the lifetime of the WSNs. By using DaT protocol, the number of data sent to the sink is reduced inside each sensor device by choosing the best data that represent the whole collected data in every period instead of transmitting all the data to the sink. The grouping method based Modified k-Nearest Neighbour is applied to classify the data of the whole period into classes. Every class includes the most similar data. Once a predefined number of initial classes are formed, the similar classes are then combined to further reduce the data size. DaT protocol is simulated using OMNeT++ network simulator on the network topology of the Intel Berkeley Research Lab (Prakash et al. 2009) real-world sensors dataset. The achieved results show the efficiency of the proposed DaT protocol in reducing the data redundancy and save energy thus enhancing the WSN lifetime while maintaining a suitable level of data accuracy.

3   The Proposed DaT Protocol

This section describes the proposed Data Transmission (DaT) Protocol for Reducing the energy consumption in Wireless Sensor Networks. Figure 1 shows the proposed Data Transmission (DaT) protocol.

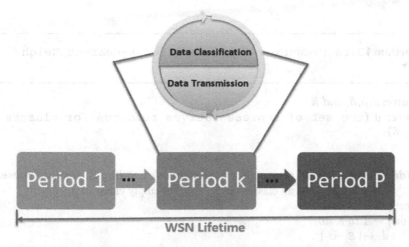

Fig. 1. Data Transmission (DaT) protocol.

DaT protocol partitions the lifespan of the WSN into periods. Two stages are found in every period. The first stage is the data classification in which the collected sensed data is classified into several classes based on the Modified k-Nearest Neighbour technique. Every class includes the most similar sensed data. Algorithm 1 shows the data reduction based modified k-Nearest Neighbour technique. DaT protocol selects the best representative reading from each class to send it to the sink rather than sending the whole data of each class. In order to further reduce the volume of data transmitted from sensor nodes to the sink node, the similar classes are merged into one class and chooses the best reading that represents all the readings in the new merged class. In the proposed DaT protocol, the period of time can be varied to ensure a certain level of data reduction as well as a desired level of data accuracy. Table 1 gives some parameters that are used in our proposed algorithm.

Table 1. Some parameters that are used in our proposed algorithm

Parameter	Meaning
S	The set of sensed data reading per period
ρ	The number of readings per period
δ	The difference threshold defined by the application
K	The number of classes
μ	The set of representative data readings
C^k	The class k
$Length(C^i)$	The number of readings in class i
E_{elec}	The power dissipation by the radio to operate the transmitter or receiver circuits
β_{amp}	The amplifier of transmitter

Algorithm 1 Data Reduction based modified k-Nearest Neighbour

---

Require: $S, \rho, \delta,$ and K
Ensure: μ (the set of representatives readings for classes
k=1...K)

1:
Divide S into K subsets, then select randomly one reading c_k foreach subset
2: $C^k \leftarrow \emptyset \; \forall \; k \in K$ // Set all classes (k=1, ..., K) to empty class
3: *for* $i \leftarrow 1$ *to* ρ *do*
4: *for* $j \leftarrow 1$ *to* K *do*
5:     $d_j \leftarrow |S_i - c_j|$
6:   *endfor*
7:   $k \leftarrow IndexofMin(d_j)$
8:   $C^k \leftarrow C^k \cup S_i$ // Assign Si to the nearest class C^k
9: *endfor*
10: *for* $k \leftarrow 1$ *to* K *do*
11:     $\mu_k \leftarrow \dfrac{\sum_{i=1}^{Length(C^k)} c_i^k}{Length(C^k)}$
12: *endfor*
13: *for* $i \leftarrow 1$ *to* $K - 1$ *do* // Merge the similar classes
14: *for* $j \leftarrow i + 1$ *to* K *do*
15:     *if* $(|\mu_i - \mu_j| \leq \delta)$*then*
16:       $C^i \leftarrow C^i \cup C^j$
17: *Remove* C^j *from* μ_j
18:       $\mu_i \leftarrow \dfrac{\sum_{l=1}^{Length(C^i)} c_i^i}{Length(C^i)}$
19:       $K \leftarrow K - 1$
20:   *endif*
21: *endfor*
22: *endfor*
23: *return* μ

In Algorithm 1, the set of sensed data readings is virtually divided into subsets. The idea is to select from each subset an initial reading to form the centroid of a new class. These centroid readings are selected randomly. However, each of the remaining data readings in the whole sensed data set is compared to these centroid readings. The data reading will belong to the nearest centroid reading. Once the nearest centroid reading is determined, the data reading is assigned to the class that is represented by that nearest centroid reading. The process continues till all data readings are labeled with the class whose centroid reading is nearest to them. In the data transmission stage, each sensor node will send the reduced vector that resulted from the first stage to the sink. This reduced vector (μ) represents the best representative readings of all the classes. Finally, the classes with the newly-calculated centroid reading are compared to each other, and the similar classes are merged together to produce an even smaller number of classes then fewer number of centroid readings. Finally, Algorithm 1 needs $O(\text{Max}(\rho \cdot K, K2 \cdot \text{Length}(Ci)))$ of time complexity. It requires space complexity equal to $O(\text{Max}(\rho, K \cdot \text{Length}(Ci)))$. The message complexity is $O(\mu)$. The algorithm has the advantage of being computationally undemanding which translates into less processing and thus reduced level of energy consumption. In the second stage, the best representative readings are selected from every class to send it to the sink.

3.1 Illustrative Example

In order to demonstrate the algorithm, a real data of sensor nodes from the Intel Berkeley Lab Datasets are used. For the sake of simplicity, $\rho, \delta, \wedge K$ are set to 10, 0.05, 3 respectively. Let the sensed data reading S = {19.9884, 19.3024, 19.1652, 19.175, 19.1456, 19.1652, 19.1652, 19.1456, 19.1456, 19.1456}, the algorithm can be applied as follow:

A. Divide the data readings set virtually into three subsets. Therefore, the three subsets (classes) are

Subset 1 (class 1)	Subset 2 (class 2)	Subset 3 (class 3)
19.9884, 19.3024, 19.1652, 19.175	19.1456, 19.1652, 19.1652	19.1358, 19.1456, 19.1456

B. One reading is selected at random from each subset and called centroid reading. C = {19.9884,19.1456,19.1358}

C. For each of the other seven remaining readings, calculate the difference between the data reading and the centroid reading and then attach the data reading to the nearest centroid reading.

Sample	19.9884	19.1456	19.1358	Assigned class
19.9884	0	0.8428	0.8526	1
19.3024	0.686	0.1568	0.1666	2
19.1652	0.8232	0.0196	0.0294	2
19.175	0.8134	0.0294	0.0392	2
19.1456	0.8428	0	0.0098	2
19.1652	0.8232	0.0196	0.0294	2
19.1652	0.8232	0.0196	0.0294	2
19.1456	0.8428	0	0.0098	2
19.1456	0.8428	0	0.0098	2
19.1358	0.8526	0.0098	0	3

D. Hence, eight of the 10 data readings are labeled as class 2, while class 1 and class 2 contain only one data reading for each class. Therefore, the three classes can be seen as follows

> Class 1: 19.9884
> Class 2: 19.3024, 19.1652, 19.175, 19.1456, 19.1652, 19.1652, 19.1456, 19.1456
> Class 3: 19.1358

E. for each of the three newly-formed classes:
   a. calculates the centroid of the newly-formed class by taking the mean of its data reading.

Class 1 centroid	Class 2 centroid	Class 3 centroid
19.9884	19.1762	19.1358

   b. Compare each class with the rest and merge if less than or equal to a certain threshold.

Class 1 to class 2	Class 1 to class 3	Class 2 to class 3
0.81	0.85	0.0404

Since the distance between class 2 to class 3 is less than 0.05, the two classes will be merged into one class. Hence, the final classes will be as follows:

> Class 1: 19.9884 its centroid is 19.9884
> Class 2: 19.3024, 19.1652, 19.175, 19.1456, 19.1652, 19.1652, 19.1456, 19.1456, 19.1358. Its centroid is 19.1717

F. Finally, the representative readings that will be sent to the sink are 19.9884 and 19.1717.

4   Simulation Results and Analysis

The main objective of the DaT protocol is to reduce the communication cost by eliminating the redundant sensed data from the collected data of sensor nodes before sending it to the sink. This can save the power of the sensor nodes and enhance the WSNs lifetime. This section is focused on the evaluation of the proposed DaT protocol using OMNeT++ network simulator [22]. DaT is based on real sensed data from a sensor network deployed in the Intel Berkeley research lab [21]. This dataset consists of 2.3million sensed data readings (like humidity, temperature, light and voltage values) and gathered by 54 nodes of type Mica2dot. One reading every 31 s is captured by the sensor node. Table 2 shows the values of the simulation parameters in DaT protocol.

Table 2. Parameters values of simulation.

Parameter	Value
WSN size	54 nodes
ρ	20, 50, and 100 data readings
δ	0.03, 0.05, and 0.07
K	$\rho/2$
E_{elec}	50 nJ/bit
β_{amp}	100 pJ/bit/m2

In order to show the effectiveness of DaT protocol, it is compared with two recent exciting methods in the literature like PFF [19] and ATP [18]. Several performance criteria are used to evaluate the proposed DaT protocol such as percentage of data after aggregation, energy consumption, and data accuracy.

The sensor nodes used the same energy consumption model discussed in [23]. The cost of sending is computed for a f – bits packet and for a distance d is calculated as follows

$$E_{Send}(f, d) = E_{elec} * f + \beta_{amp} * f * d^2. \tag{1}$$

The consumed energy needed to receive f – bits is computed as follow

$$E_{Receive}(m, d) = E_{elec} * f. \tag{2}$$

Where E_send and E_receive refer to the energy consumption required for sending and receiving f – bits respectively. Therefore, the communication cost represents both E_send and E_receive inside the sensor node. the communication cost can be increased when the E_send and the E_receive are increased.

4.1 Percentage of Data After Applying Data Reduction

This experiment shows that the sensor node can decrease the volume of transmitted data to the sink. Figure 2 illustrates the percentage of data after data reduction for various volumes of data. The results explain that DaT protocol can decrease from 51% and up to 68% and from 89% and up to 93% the size of transmitted data from every node in comparison with ATP and PFF respectively. It can be seen that the data reduction increases when the difference threshold increases due to the increase in the similarity among the sensed data. Furthermore, the data is further reduced when the data volume increases. We can see that PFF did not remove the redundant data at the sensor node, therefore it sends the whole sensed data to the sink.

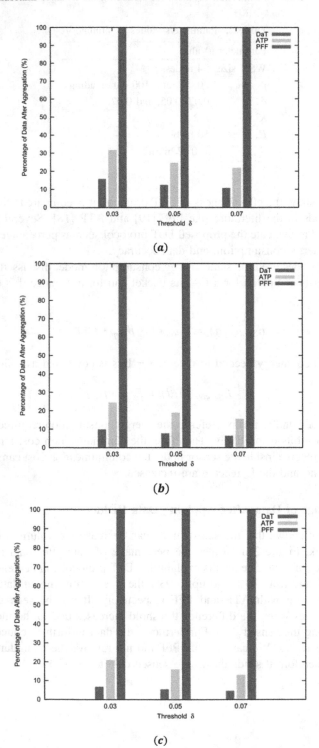

Fig. 2. Percentage of data after applying data reduction. (*a*) $\rho = \mathbf{20}$, (*b*) $\rho = 50$, *and* (*c*) $\rho = 100$.

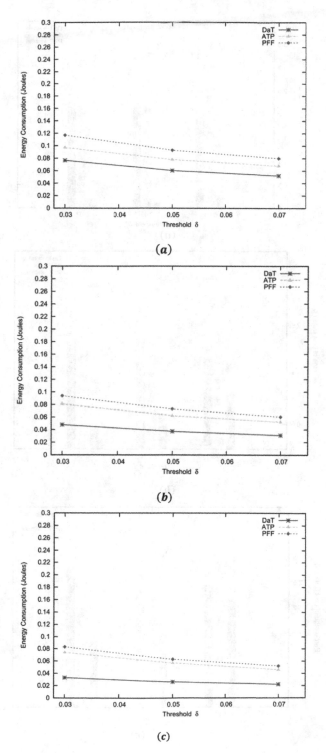

Fig. 3. Energy consumption. (*a*) $\rho = 20$, (*b*) $\rho = 50$, *and* (*c*) $\rho = 100$.

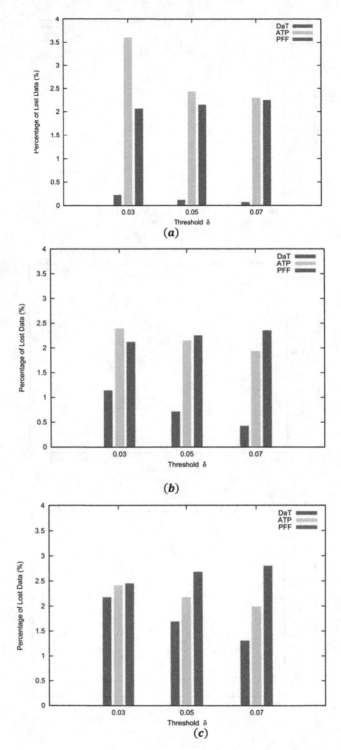

Fig. 4. Percentage of lost data. (*a*) $\rho = 20$, (*b*) $\rho = 50$, *and* (*c*) $\rho = 100$.

4.2   Energy Consumption

The whole purpose of data reduction is to reduce the consumed energy by the nodes and thus reduce the communication cost in WSNs. Therefore, it is natural that the energy saving is measured to serve as an indicator of the algorithm performance. Figure 3 explains the power consumption for various data volumes. As shown in Fig. 3, DaT protocol conserve the power of the sensor device from 23% and up to 56% and from 35% and up to 60% in comparison with ATP and PFF respectively. It can be seen that DaT protocol efficiently save the power due to eliminating the redundant data after applying the modified k-Nearest Neighbour technique.

4.3   Percentage of Lost Data

The data loss ratio can be considered as a measure for data accuracy. It refers to the number of lost data readings after receiving the sensed data readings of the sensor nodes by the sink. The conducted results are demonstrated in Fig. 4 and it can be shown that DaT protocol improves the accuracy of data from 10% and up to 97% and from 22% and up to 97% in comparison with ATP and PFF respectively. It can be seen from Fig. 4 the effect of data size on the level of lost data, whereas the size of data increases the percentage of lost data increases.

5   Conclusions

In this article, we introduced a Data Transmission (DaT)protocol in sensor networks. Our protocol is implemented locally at every sensor node and it is appropriate to the restricted resources of sensor networks. DaT is periodic and consists of two stages. Classification of the data of the period into classes based on the modified K-nearest neighbour technique, merge the similar classes are performed in the first stage. The best representative readings are sent to the sink node in the transmission stage. DaT is evaluated based on real data of the Intel Lab sensor network using OMNeT++ network simulator. The conducted results are compared with the results of two existing methods (ATP and PFF). DaT outperforms the two algorithms in terms of data reduction ratio and hence the energy saving ratio. Our DaT protocol has a better level of accuracy compared to ATP and PFF. In future, we plan to extend our work to take into account the spatial correlation among the sensor nodes to further reduce the redundant data before sending it to the sink. Furthermore, the data reduction method can be implemented within a full distributed protocol where each sensor will decide to send or not based on the cooperation with its neighbours.

References

1. Noel, A., et al.: Structural health monitoring using wireless sensor networks: a comprehensive survey. IEEE Commun. Surv. Tutor. (2016). https://doi.org/10.1109/comst.2017.2691551
2. Keswani, K., Bhaskar, A.: Wireless sensor networks. In: International Conference on Futuristic Trends in Engineering, Science, Humanities, and Technology, Gwalior, 23–24 January 2016 (2016)

3. Dias, G.M., Bellalta, B., Oechsner, S.: The impact of dual prediction schemes on the reduction of the number of transmissions in sensor networks. Comput. Commun. **112**, 58–72 (2017)
4. Zhong, H., et al.: An efficient and secure recoverable data aggregation scheme for heterogeneous wireless sensor networks. J. Parallel Distrib. Comput. **111**, 1–12 (2017)
5. Bramas, Q., Tixeuil, S.: The complexity of data aggregation in static and dynamic wireless sensor networks. Inf. Comput. **255**, 369–383 (2017)
6. Nguyen, N.-T., et al.: On maximizing the lifetime for data aggregation in wireless sensor networks using virtual data aggregation trees. Comput. Netw. **105**, 99–110 (2016)
7. Brahmi, I.H., et al.: A spatial correlation aware scheme for efficient data aggregation in wireless sensor networks. In: 40th Annual IEEE Conference on Local Computer Networks, pp. 847–854 (2015)
8. Jellali, Z., Atallah, L.N., Cherif, S.: Linear prediction for data compression and recovery enhancement in wireless sensors networks. In: IWCMC Conference 2016, pp. 779–783 (2016)
9. Sharma, R.: A data compression application for wireless sensor networks using LTC algorithm. In: 2015 IEEE International Conference on Electro/Information Technology (EIT) (2015)
10. Galhotra, K., Kaur, K.: Extending IAMCTD using data fusion and lossless data compression for UWSN. In: 1st International Conference on Futuristic trend in Computational Analysis and Knowledge Management, pp. 526–531 (2015)
11. Srisooksai, T., et al.: Practical data compression in wireless sensor networks: a survey. J. Netw. Comput. Appl. **35**, 37–59 (2012)
12. Tan, L., Wu, M.: Data reduction in wireless sensor networks: a hierarchical LMS prediction approach. IEEE Sens. J. **16**(6), 1708–1715 (2016)
13. Arbi, I.B., Derbel, F., Strakosch, F.: Forecasting methods to reduce energy consumption in WSN. In: IEEE International Conference on Instrumentation and Measurement Technology (I2MTC) (2017)
14. Ganjewar, P.D., Barani S., Wagh, S.J.: Data reduction using incremental Naïve Bayes Prediction (INBP) in WSN. In: 2015 International Conference on Information Processing (ICIP). Vishwakarma Institute of Technology (2015)
15. Ganjewar, P.D., Wagh, S.J., Barani, S.: Threshold based data reduction technique (TBDRT) for minimization of energy consumption in WSN. In: 2015 International Conference on Energy Systems and Applications (ICESA 2015), Pune, India (2015)
16. Jabera, A., et al.: Reducing the data transmission in sensor networks through Kruskal-Wallis Model. In: 2017 Tenth IEEE International Workshop on Selected Topics in Mobile and Wireless Computing (2017)
17. Harb, H., Idrees, A.K., Jaber, A., Makhoul, A., Zahwe, O., Taam, M.A.: Wireless sensor networks: a big data source in internet of things. Int. J. Sens. Wirel. Commun. Control **7**(2), 93–109 (2017)
18. Harb, H., Makhoul, A., Couturier, R., Medlej, M.: ATP: an aggregation and transmission protocol for conserving energy in periodic sensor networks. In: 2015 IEEE 24th International Conference on Enabling Technologies: Infrastructure for Collaborative Enterprises (WETICE), pp. 134–139. IEEE (2015)
19. Bahi, J.M., Makhoul, A., Medlej, M.: A two tiers data aggregation scheme for periodic sensor networks. Adhoc Sens. Wirel. Netw. **21**(1) (2014)
20. Idrees, A.K., Al-Yaseen, W.L., Taam, M.A., Zahwe, O.: Distributed data aggregation based modified K-means technique for energy conservation in periodic wireless sensor networks. In: IEEE Middle East and North Africa Communications Conference (MENACOMM), Jounieh, Lebanon, pp. 1–6. IEEE (2018)

21. Madden, S.: Intel Berkeley Research Lab (2004). http://db.csail.mit.edu/labdata/labdata.html
22. Varga, A.: nesC language manual (2003). https://github.com/tinyos/nesc/blob/master/doc/-ref.pdf?raw=true
23. Heinzelman, W.R., Chandrakasan, A., Balakrishnan, H.: Energy-efficient communication protocol for wireless microsensor networks. In: Proceedings of the 33rd Annual Hawaii International Conference on System Sciences, p. 10-pp. IEEE (2000)
24. Idrees, A.K., Deschinkel, K., Salomon, M., Couturier, R.: Multiround distributed lifetime coverage optimization protocol in wireless sensor networks. J. Supercomput. **74**(5), 1949–1972 (2017)
25. Idrees, A.K., Deschinkel, K., Salomon, M., Couturier, R.: Perimeter-based coverage optimization to improve lifetime in wireless sensor networks. Eng. Optimiz. **48**(11), 1951–1972 (2016)
26. Hussein, W., Idrees, A.K.: Sensor activity scheduling protocol for lifetime prolongation in wireless sensor networks. Kurd. J. Appl. Res. **2**(3), 7–13 (2017)
27. Idrees, A.K., Deschinkel, K., Salomon, M., Couturier, R.: Distributed lifetime coverage optimization protocol in wireless sensor networks. J. Supercomput. **71**(12), 4578–4593 (2015)
28. Idrees, A.K., Harb, H., Jaber, A., Zahwe, O., Taam, M.A.: Adaptive distributed energy-saving data gathering technique for wireless sensor networks. In: Wireless and Mobile Computing, Networking and Communications (WiMob), pp. 55–62, October 2017. IEEE (2017)
29. Idrees, A.K., Al-Qurabat, A.K.M.: Distributed adaptive data collection protocol for improving lifetime in periodic sensor networks. IAENG Int. J. Comput. Sci. **44**(3) (2017)
30. Rault, T., Bouabdallah, A., Challal, Y.: Energy efficiency in wireless sensor networks: a top-down survey. Comput. Netw. **67**, 104–122 (2014)
31. Soni, S.K., Chand, N., Singh, D.P.: Reducing the data transmission in WSNs using time series prediction model. In: 2012 IEEE International Conference on Signal Processing, Computing and Control (ISPCC), WaknaghatSolan, India, pp. 1–5. IEEE (2012)

A Comprehensive Review for RPL Routing Protocol in Low Power and Lossy Networks

Athraa J. H. Witwit(iD) and Ali Kadhum Idrees(✉)(iD)

Department of Computer Science, University of Babylon, Babylon, Iraq
athraajawad87@gmail.com, ali.idrees@uobabylon.edu.iq

Abstract. Low Power and Lossy Networks (LLNs) are composed of a large number of nodes which are characterized by limited resources (like energy, memory, processing power, and bandwidth). These nodes are interconnected by lossy links. Therefore, they support low data rates. This leads to an unstable state with relatively low rates of packet delivery. In the beginning 2008, the Internet Engineering team (IETF) working team, named ROLL, investigates to use the existing routing protocol for LLNs, but they found that these protocols were not sufficiently appropriate for LLNs. Due to the important role of the LLNs in the Internet of Things (IoTs), ROLL standardizes a routing solution called IPv6 for LLNs. Therefore, ROLL began to design a new protocol based on IPv6 Routing Protocol for Low Power and Lossy Networks named (RPL). It is an IPv6 routing protocol specifically designed for low power and lossy networks (LLN) compliant with the 6LoWPAN protocol. RPL has gained a lot of maturities. It is attracting increasing interest in the research community. RPL is flexible in building the topology. It constructs a topology proactively. Due to the absence of a fully comprehensive review about RPL drive us to introduce this article. Furthermore, this article investigates the most related studies obtained about RPL routing protocol that concern to its implementation, performance, applications evaluation, and improvement. An open research challenges on the RPL design are pointed out. Finally, this survey can support researchers to further understand the RPL and participate to further improve it in the future research works.

Keywords: Internet of Things · Low Power and Lossy Networks
ROLL · IETF · IPv6 · RPL routing protocol

1 Introduction

Recently the world of networking is entering in a new era, named the era of Internet of Things (IoTs) [49]. IoT definition is a large number of physical objects such as cars, homes, devices with embedded microprocessors, sensors, etc. This kind of network authorizes the objects to gather and exchange the data. It makes the computer system and physical world integrated directly [15]. The IoTs involves several technologies like smart webs, smart houses, smart cities, and intelligent transportation. The main goal of these network technologies is to enhance the efficiency, reliability, and price [28]. Wireless Sensor Networks (WSNs) represent one of the big data providers in IoTs [51, 52]. Sensor nodes are the basic element in the IoT. Nodes have limited computation, bandwidth, and

© Springer Nature Switzerland AG 2018
S. O. Al-mamory et al. (Eds.): NTICT 2018, CCIS 938, pp. 50–66, 2018.
https://doi.org/10.1007/978-3-030-01653-1_4

battery life [47, 48, 50]. The nodes communicate with the Internet through a border router which is a single central point in the network [11]. There should be messages move among nodes before reaching to the central point because sometimes the size of the WSNs might be very big. All the interactions between the border router and the nodes should be done by wireless. This complexity of the network reveals an important problem that should be resolved to make the network effective and steady for a long-time period. In networking, the functions of processing and delivering of packets from one node to another in a network are done by routing protocols in the network layer. Routing protocol gives the decisions for packet transmission toward its destination. IETF (WG) had introduced RPL, the IPV6 routing protocol for LLNs. RPL has contributed to the progress of tiny, embedded networking devices for IOT. The main characteristic of RPL is that it is produced for lossy links networks that reveal a high rate of packet error and link power disconnect. RPL is a favorable stabilized solution for LLNs even though it is still a RFC [27]. In Fig. (1), a lot of endpoints forms LLNs deeply and rooted at LBR. LLN is connected to a wide area network(WAN) through LBR. The WAN consist of either private IP-network or public internet [31].

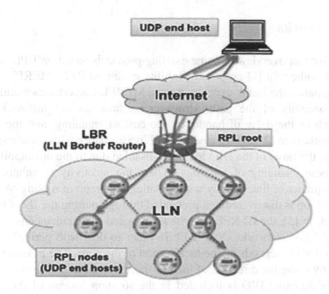

Fig. 1. An IoT multihop scenario.

For the best of our knowledge, this comprehensive survey of RBL protocol is the first inclusive research.

Topology is built by RPL, where it came from the concept of Directed Acyclic Graph (DAG), named Destination Oriented DAGs (DODAGS). RPL depends on DAG. The DAG look likes the structure of a tree, where default paths among nodes in the LLN are determined by this structure. The node in the DAG can belong to more than one parent in comparison with a traditional tree. Furthermore, the nodes are organized by RPL as a DODAGs are looking like to the topology-based cluster tree. In DODAG architecture, nodes associated not only with its parent (with higher rank) but also to other sibling nodes. The routing protocol works by allocating a rank for each node. Therefore, when node far from the border node, then node's rank increases [2, 21]. Each node that does not have the same constraints and bonds as the sensor node is called border router. This kind of node will not collect its own data also it will not be connected to a power that is why it does not count as a power usage. The border router's function is an internet mediator and gathers the received information from the WSN nodes [3, 4]. Moreover, sending of the updates to the other nodes are controlled by the border router. There are two directions for the RPL work, from the border router: upward routing transferring packets from the leaf node towards the border router, and then downward routing sending packets from the border router to any node [23].

This comprehensive survey is partitioned into two main parts: The first part introduced in Sects. 1–5 which exhibit an introduction, related literature, scientific background of RPL and its main features, RPL control messages, and DODAG construction, Repairing, and maintenance. The second part (i.e. Sects. 6–8) provides a comprehensive review of the latest advancement in the RPL research works, RPL Routing Challenges, and discussions and further analysis.

2 Related Works

This section gives an overview of some existing protocols based on RPL standard like in [1–27]. The authors in [1] proposed Mobility enhanced RPL (MERPL) to improve the RPL and balance the traffic over the routes. MERPL outperforms standard RPL in terms of the stability of the path. However, authors do not utilize some major parameters such as the delay of handover, the cost of signaling, and the energy dissipation. The authors in [2] investigated the RPL performance with some modifications. In the first step, the timer of the DIO trickle is disabled due to the unsuitability of using it in the periodic changing of topology. Then, the reactivity is enhanced through evaluating the quality of link directly with updating the graph of routing. After that, the problem of the loop is discovered and prevented through putting the ID of the parent in the DIO packet. In [3], the GI-RPL scheme is proposed as an enhancement to the RPL in Vehicular Ad-hoc Networks (VANET). In order to deal with periodic changing in the topology, GI-RPL depends on the localization of the node. The position of the node is determined by using the direction of the car and the distance to the sink. In addition, the duration of adaptive DIO is included in the solution instead of the timer of the trickle to increase the performance. an enhancement is introduced by these solutions to the packet submission rate and the message delay. Nevertheless, the constraint of power is not considered due to the characteristics of VANETs.

The work in [4], the authors proposed a modification to the cluster-directed acyclic into a cluster-directed acyclic graph (DAG) to improve the redundancy of topology at the MAC layer because that structure was followed by IEEE 802.14.2 which lead to network partitions even after single link or node failure. In [5], a way for transmitting the sensed data from through a single path from sensors to root is proposed. Nevertheless, depending on the unreliability of wireless links and the constraints of sensor nodes, a single path is not an effective strategy to cope with the demands of the performance of different applications. Therefore, the authors suggest three kinds of multipath methods depending on RPL (Energy Load Balancing (ELB), Fast Local Repair (FLR), and their integration ELB-FLR. In [6], the communication path is very important for RPL because it passes to the center of the router that might give near optimal paths making no uniqueness for the application-driven extension to the proposed RPL. This enables the increase in the WSN lifetime limiting the functions of the network routing and forwarding, especially, the node that uses the similar applications.

In [8], a novel method named Co-RPL is proposed. The Corona scheme is used by this solution positioning the nodes according to the position of the DAG root to discover the movement of these nodes. Nevertheless, the problem of mobility (e.g., MN disconnection) does not be solved in this solution. The work in [16] introduced a new objective function to balance the number of overloaded children nodes to ensure the maximization of the node lifetime. The implementation of the mode (OF), they modified the form of DIO layout. Using new technology. In [18], the authors extend the RPL from the mobility point of view. The presented schemes are analyzed to observe the impact on the LLNS requirements. The authors in [9] check the RPL performance in the density of the medium using the two objective functions in various topologies.

The research in [20] focused on the requirement for an intermediate layer between network and mac layers to get rid of the negative impact on the redundant paths numbers and their quality and overall performance of the routing protocol. In [21], authors suggested a hybrid RPL routing approach based on power-saving cluster-parent named HECRPL. The main goal of this protocol is to perform simultaneously reliability and energy-saving. The work in [22], they presented an enhanced version of RPL called enhanced-RBL to mitigate the problem of strong limitation of the preferred parent nodes. The node is allowed to distribute prefixes belonging to its subnetwork among multiple parents. In [25], authors proposed MRBL protocol to deal with mobility based on a proactive approach. In spite of this scheme presented a good contribution and succeeded to improve some needed performances, the enhancement is still needed.

The research in [26] displays a study of RPL with the latest updated version. For example, point to point RPL (P2P-RBL), assessment, performance, challenges, research defiance, and the RPL visualized opportunities. The authors in [27] proposed Objective Function Based on Fuzzy logic (OF-FL). This is a new function that gets rid of the standard objective functions restriction which are developed for RPL through taking into account important nodes and links metrics like hop-counts, end-end delay, Link Quality Level (LQL), and Expected Transmission Count (ETX). In [28], they proposed a new solution which aims to achieve distributed monitoring with minimal computational complexity. The main objective is to increase robustness in IoT via monitoring the links in the destination-oriented directed acyclic graph (DODAG) constructed by RPL.

3   Background

RPL is a Distance Vector (DV) protocol-based source routing which is developed for working upon several link layer techniques which includes -IEEE 802.15.4 -PHY and MAC layers. It aims at gathering-based networks in which a node transmits the values to a gathering point periodically, and also send point-to-multipoint traffic from the center point to the devices within the LLN. In addition, point-to-point traffic is supported also in RPL.

RPL has the ability to exemplifies a routing solution for LLNs, which refers to very limited resources networks from the power, processing, and bandwidth making the networks in risk of losing packets [32, 33]. In fact, RPL has developed to support the nodes have limited resources. Two principal characteristics are considered by RPL-enabled LLNs: (i) it has low data rate, and (ii) It is exposed to high error rates that lead to decrease the throughput.

The long inaccessibility time and high Bit Error Rate (BER) are features of the lossy link which influence the design of routing protocol. RPL is designed to provide alternative paths when the default paths are not found and to be adaptive for the condition of the network [31]. RPL depends on the concept of topology that named Directed Acyclic Graphs (DAGs). The tree structure is formed by the DAG that determines among the nodes of the LLN, the default paths. The DAG is more than a normal tree because every node might be connected to multiple parent nodes in the structure of DAG, in contrary to the typical trees whereas just a single parent is to be allowed.

RPL sorts the nodes as Destination-Oriented DAGs (DODAGs), the place of sinks or the ones giving a route to the Internet (i. e, gateways) behave as the DAGs root. A network might include one or more than one DODAG, that makes an RPL instance modified by a unique ID, named RPL Instance ID. A network can operate a group of RPL instances simultaneously. These instances are naturally independent. A node is able to connect to multiple RPL instances, but it should only relate to a single DODAG within every instance. Some DAG type strategies in the network layer can be investigated in [54, 55]. Figure 2 illustrates RPL instances with various DODAGs.

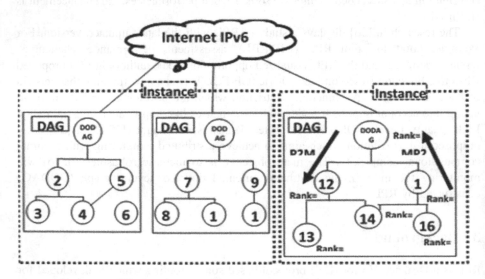

Fig. 2. RPL instances with various DODAGs [27].

The hierarchical nature of the RPL-based network topology makes self-organization of nodes based on the relationship of parent-to-child. The mesh topology is supported by RPL due to enabling the routing by brothers rather than parents and children. Great flexibility from the topology management and routing point of view is provided due to the combination of the mesh and hierarchical topologies. Table 1 outlines the principal characteristics of the RPL protocol.

Table 1. Outlines the principal characteristics of the RPL protocol

Feature	Description
Target network	LLN; IPv6/6 low PAN network
Routing type	Source-routing, Distance – Vector
Topology	Mesh/hierarchical based on DAGs
Traffic flows	MP2P, P2MP and P2P
Message update	Trickle timer
Control message	DIO, DAO, DIS
Neighbor discovery	Like IPV6ND mechanisms
Transmission	Unicast and multicast
Metrics and constraints	Dynamic, based on OF and Rank
Modes	Storing and non-Storing

4   DODAG Specification

In this Section, we display the principal messages and mechanisms provided in RPL as defined in IETF drafts.

4.1   RPL Control Messages

The new ICMPv6 control messages types are defined by RPL messages, whose structure is described in Fig. 3 [37].

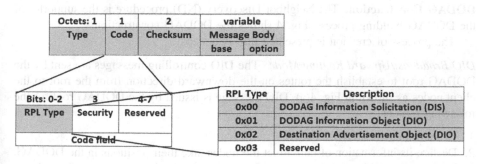

Fig. 3. RPL control message.

The control message of RPL consists of (i) an ICMPv6 header, which composed of three fields: Type, Code, and Checksum, (ii) Body of the message including message contents and some options. The ICMPv6 control packet is determined by the type field. In the case of RPL, it is set to 155. The type of RPL control packet is distinguished by the Code field. Four codes are currently identified [31]:

DODAG Information Solicitation (DIS) is designated by 001. The DIS message is primarily operated to obtain a DODAGs Information Object (DIO) from the RPL node. The DIS might be used to examine nearby node in neighboring DODAG. The present DIS message formation has no-particular fields and flags for the next usage.

DODAG Information Object (DIO). DIO message is designated at 0x01. It is assigned by the DODAG root to construct a new DAG and then transmitted to a group of nodes through the structure of the DODAG. The DIO message holds related network data that permits to a node to find the RPL instance, preserve the DODAG, discover its configuration parameters, choose a DODAG parent list, and preserve the DODAG.

Destination Advertisement Object (DAO). The message of DAO is assigned to 002 and is utilized to deliver reversed route data to register each visited node over the particular upward route. Every node sends DAO message except the DODAG root, to propagate the prefixes and routing tables of children to their parents. After that, the complete route will be established after passing the DAO message over the route from a certain node to the DODAG root.

Destination Advertisement Object (DAO-ACK). The DAO-ACK message is delivered individually from a DAO receiver (a DODAG root or DAO parent) as a reply to the received unicast DAO message. It transfers data regard Status, DAO Sequence, and RPL Instance ID, which indicate the completion. The codes of status are not defined yet, but the codes larger than 128 suggest that the parent is rejected and another parent should be chosen as an alternate.

4.2   DODAG Construction, Repairing, and Maintenance

In this section, we introduce the main steps for building DODAG tree and how to repair it in the case of failure as well as the way of maintaining this construction up to date.

DODAG Construction. The Neighbor Discovery (ND) procedure is the supporter of the DODAG building process, Fig. 4 shows the DODAG construction.

The process of creation is presented into two steps:

DIO Broadcast/Upward Routing Mode. The DIO controlling messages are sent by the DODAG root to establish the routes on the downward direction from the root to the client nodes as shown in Fig. 4(a). DIO message is issued by the DODAG root for the following reasons:

1. Declare their own DODAGID.
2. Declare its information of rank to let the nodes take their positions in the DODAG.
3. Declare the Objective Functions distinguished by the Objective Code Point (OCP) inside the configuration option fields of DIO message.

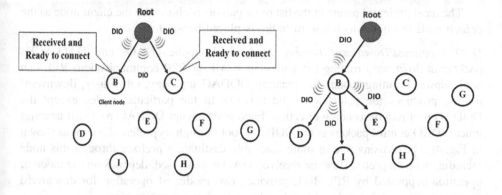

(a)   (b)

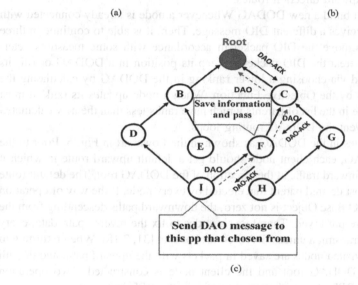

(c)

Fig. 4. DODAG construction: (a) DIO broadcast, (b) DIO Packet update broadcast, and (c) DAO unicast packet.

This particular message would be received by the client node, which can be a node ready to join or an already joined node. When a node decides to join in the DODAG, then it:

1. Inserts the address of DIO sender to its own parent list.
2. Evaluates its ranks in accordance with the Objectives Function determined in the OCP filed, where node's rank is greater than all its parents.
3. Sends the DIO message notification along with the information of the updated rank to the DODAG root and all the nodes connected with it as shown in Fig. 4(b).

The most preferred parent in the list of its parents is chosen by the client node as the default node through which inward traffic is placed forward.

DAO Broadcast/Downward Routing Mode. RPL applies DAOs and DAO-ACKs packets in downward routing for achieving P2MP or P2P communication. RPL can apply upward routing mode after creating DODAG topology. Obviously, downward routing produces the paths down the network to the particular nodes except the DODAG root in a descending direction. Each node except DODAG root will transmit unicast DAO control packets to the DODAG root through its preferred parent as shown in Fig. 4(c) advertising one or more reachable destination prefixes through this node including its own prefix. How the received DAO is processed, depends on the mode of operation supported by RPL. RPL provides two modes of operation for downward routes, namely, storing (table-driven) and non-storing (source-routing).

The client nodes will transmit unicast DAO control packets to the DODAG root in order to build the upward direction routes.

Therefore, it can build a new DODAG Whenever a node is already connected with DODAG then it receives a different DIO message. Then, it is able to continue in three different ways (i) remove the DIO packet in accordance with some measures determined by RPL, (ii) treat the DIO packet to keep its position in a DODAG or (iii) its location is enhanced via choosing a lower ranking in the DODAG by calculating the cost of path chosen by the Objective Function. When a node updates its rank, it must leave out every node in the list of parents which have ranks less than the new calculated node's rank to prevent the DODAG routing loops.

The node operates in the DODAG as shown in the flowchart in Fig. 5. Prior to the making of a DODAG, each client node should get a default upward route in which it could broadcast its inward traffic at the direction of the DODAG root. The default route is created by the most desired parent that elected by every node. If the way of operation flag mode in the DIO Base Object is not zero, the downward paths descending from the root to the nodes are preserved. Therefore, in order to fix the reverse path data, every client node should transmit a unicast DAO control packet [31, 2.18]. When returning to the DODAG root, visited nodes are saved in packets with the upward path, and the full route between the DODAG root and the client node is constructed. Two operation modes are used by RPL to save downward paths in the RPL instance:

1. Storing mode
   A DAO packet is delivered in unicast way from a particular child to a certain parent that could save DAO packets sent from their child prior to sending the new DAO packet with the total information of reachability to its parent. A multicast mode could be activated or deactivated depending on the storing method.
2. Non-storing mode
   In the non-storing technique, DAO packets are transmitted to the root of DODAG and it is not stored by the intermediate parents. It is only insert its address to the reverse route stacks in the transmitted DAO packet, then sends it to its parent.

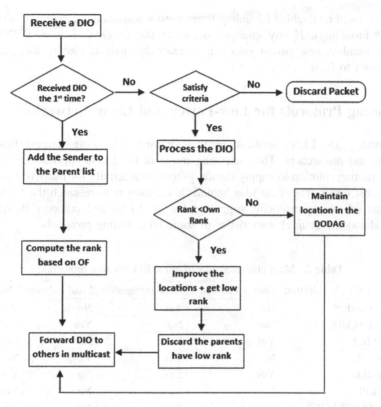

Fig. 5. Node operation in the DODAG

DODAG Repairing. When inconsistencies are discovered, a repair must be done. This repair can be performed in two ways, locally or globally. The repairing technique that made Locally to mend inconveniences like link failure or a loop discovery.

It solves the discovered inconsistency without rebuilding the DODAG from the root. For example, a new parent will be selected if the node is failed to connect with its parent. The local repairing cannot provide optimal solutions. Every once a local repairing is done, the optimal state of DODAG will diverge because a global repairing will finally be required. The construction process will be restarted by a global repairing and yield a new version number to the DODAG. The optimal construction is provided by the global repairing but this operation is costly from the network performance point of view because of the new construction of the graph that leads to increasing the overhead of control messages in the network [18].

DODAG Maintenance. To preserve the DODAG, every node produces DIO packets in a periodic way triggered with the aid of a trickle timer. The main objective of the trickle timer is to optimize the frequency of message sending relying on the network circumstances. Briefly, the frequency of transmitted packets is decreased when the network is consistent and increased in the opposite state. The timer period increases exponentially at any time the timer fires. First, the duration of the timer is initialized to

Imin, which will be doubled Idoubling times until it reaches the maximum value Imax (¼Imin * Idoubling). If any changes occur in the DODAG like new DODAGS sequence number, new parent selection, unreachable parent, routing loop, etc., the timer is reset to Imin.

5   Routing Protocols for Low-Power and Lossy Networks

In the past years, LLNs applications were having a growing interest from both industries and researchers. The increasing range of the LLNs applications needs an effective routing solution to supply the ubiquitous connectivity for massive low-power and low-cost nodes. The main idea behind this section is to research the traits of the overall set of routing protocols suggested for the LLNs and compare its qualities. Table 2 shows the main characteristics of some RPL routing protocols.

Table 2. Main characteristics of some RPL routing protocols

No	Protocols classification	Energy efficiency	Data aggregation	Load balance	Multipath
1	Rpl standard	No	Yes	No	No
2	Rpl-BMARK	No	No	Yes	No
3	P2P Rpl	Yes	Yes	No	No
4	Co-RPL	No	Yes	Yes	No
5	Qu-RPL	Yes	Yes	No	Yes
6	Ec-RPL	Yes	Yes	No	Yes
7	ENHANCED-RPL	Yes	Yes	No	Yes
8	ER-RPL	No	No	Yes	Yes
9	C-RPL	Yes	No	No	No
10	ME-RPL	No	Yes	Yes	No
11	GI-RPL	Yes	No	No	Yes
12	MoMoRo	No	No	No	Yes
13	MRPL	Yes	Yes	Yes	No
14	BD-RPL	No	No	Yes	Yes
15	Strengthen-RPL	Yes	No	No	Yes
16	HECRPL	Yes	Yes	No	Yes

6   Routing Challenges in Low Power and Lossy Networks

RPL has developed as a prominent solution for routing in its M2M correspondences because of its perfect adaptability. The specifications of RPL routing are made to meet several needs of various LLNs applications. This research demonstrates the principal characteristics of RPL, where the essential objective is to supply readers with a complete knowledge about RPL. Several problems in RPL need to precisely addressed for ubiquitous usage of LLNs applications in the physical environment. Some main research challenges are summarized as follows.

6.1   Efficient Routing Support for Generic Traffic Patterns with Limited Memory

The applications of LLNs need a stable routing help for general data traffic, for instance, P2P, MP2P, and P2MP with heterogeneous abilities for the nodes. As previously discussed, RPL builds DODAGs in a manner that the routing help for MP2P data traffic is optimized. Nonetheless, the transmitted data of P2P and P2MP require moving over the preconstructed DAG. Therefore, the effective routing cannot be given by RPL and the delivery of data undergoes from lossy wireless links and high delay. More precisely, the communication of P2P that required by several applications of LLNs should be supported by a mediator in RPL. In the mode of non-storing, the root of DODAG represents an intermediate node. Therefore, it will be the bottleneck with the high data traffic. The P2MP data traffic is supported using the downwards paths in DODAG. Moreover, because the wireless channels asymmetric nature, the downwards paths perhaps not be the best paths. This wireless links asymmetric nature affects the performance of routing algorithms in LLNs. The protocols do not support the asymmetry of links will fail when they are applied to the asymmetric links in real networks.

6.2   Energy-Efficient Route Discovery Under Severe

In the LLNs, the best route among 2 random nodes is not supplied by default because of the restricted memory and the size of the buffer. Therefore, to support the communication of P2P, a path detection is needed. The best route is detected by P2P-RPL in a reactive routing protocol. This leads to a great overhead in dense and large networks. Throughout the path discovery, the whole nodes require co-operating to form the temporary DODAGs. This can lead to increased energy consumption. It is important to find the best route in LLNs taking to account energy saving. The reliability should be considered during protocol design.

6.3   Reliable and Energy-Efficient Routing in LLNs

Two main objectives should be considered during developing routing protocols for LLNs: reliable data delivery and energy saving. In spite of RPL supports multicast routing, it chooses an only one node as a favored parent using a certain objective function. This single route routing is not enough to be considered in the dynamic changing topology and lossy wireless channels. The packet retransmission and packets loss would be increased due to this unreliability in wireless channels. This can lead to a long time of channel occupancy and higher power consumption. In order to perform reliability and energy saving in a simultaneous way represent a difficult task in LLNs.

6.4   Congestion Detection and Control

In LLNs, the congestion takes place in both link and node levels. The limited buffer size of the node leads to overflow in the case of the heavy traffic. In addition, the congestion leads to decrease the quality of channel and extends the rates of loss. This can lead to extended delays, competition on the channel, and packet drop. Several methods are proposed to deal with congestion problem like (1) Changing the rate of traffic, (2) when congestion occurs the traffic is re-routed, (3) Dividing traffic into several routes. These methods cannot provide efficient routing in LLNs without taking into consideration the lossy nature of the LLNs. RPL design did not consider good mechanisms to deal with the congestion. Therefore, it is important to provide efficient schemes for congestion discovery and control to enhance the RPL performance.

6.5 Mobility Support

RPL is devised for fixed networks. Nevertheless, many applications in LLNs need a support of routing for mobile nodes. The dynamics changing of topology is greatly increased by mobility. It is necessary to make a modification for RPL to support the mobility in the network.

6.6 High Throughput with a Low Duty Cycling of Lower Layer

The multipath fading and interference are alleviated during design IEEE 802.15.4. It is designed to works with low energy and it represents the better solution for medium access control (MAC) layer schemes to support the M2M communications. Several characteristics in IEEE 802.15.4 like reservation of guaranteed time-slots, low consumed energy with low duty cycles and a carrier sense multiple access-collision avoidance (CSMA-CA) scheme. The dynamic duty cycling leads to low throughput, increased latency, low ratio of packet delivery. It is essential to consider the properties of the protocols of other layer during design the routing protocol. Hence, to optimize the RPL performance, the cross-layer design is needed.

6.7 Workload Balancing Strategy for RPL

The unbalanced distribution of workload is resulted from deploying the nodes randomly and heterogeneous traffic patterns in LLNs. Therefore, in the hotspot regions, nodes consume their power in fast way. During the data gathering applications, the nodes which are near the gateway will consume their power in fast way resulting in power hole problem that affect on the network performance. Hence, it is required Efficient data collection approach to mitigate the impact of this problem whilst saving the desired reliability. Furthermore, sharing the workload can treat the problem of the energy holes and extend the network lifetime.

6.8 Security Issues

In RPL, the security design is not marked. The secure connection establishment and the keys exchanging are not specified for both management and authentication in RPL. The process of joining in DODAG is not secure in RPL. The data might be lost and packets dropped. Therefore, it is easy to attack the RPL. Moreover, in LLN e-Health application, very personal and sensitive information are handled that cannot be revealed to other persons. Therefore, it is critical to develop security schemes and improve privacy saving in RPL. Therefore, in order to maintain the privacy of users, it is essential to encrypt the data and sent as cipher-text. Nonetheless, the data volume and the communication cost are increased due to the encryption that might be unsuitable for LLNs applications. Furthermore, end-to-end encryption can increase the computation overhead, consumed power, and latency. Therefore, it is required to improve security schemes which are appropriate to LLNs. Several attacks on the RPL can be occurred due to exploiting the routing rules that executed in RPL. The rank attack is one of these attacks. The malicious node in this attack does not respect the rule of the rank in RPL. This can lead to producing a bad path for transmitting the data. Some other attacks and security issues, especially on RPL such as local repair and resource-constrained attack [53].

7 Discussions and Further Analysis

This article introduces a review of the main characteristics of routing protocols in RPL. The main goal is to give to researchers a comprehensive review of RPL. In spite of the specification of RPL has obtained adequate development, the ROLL working group did not address the open research problems which are started out of the RFC specification scope. One essential open research issue that still remains open is the objective function specification. It is very important because it affects the performance of routing and the policies of route choosing. There is no more explication about how to map an Objective Function to rank computation in the current RFC specification. Furthermore, the routing metric that utilized by the objective function for optimization is not specified by RPL. Many areas of research in this scope can be defined such as the objective function specification for a particular type of applications based on their needs. For the control of topology, the selection scheme of the route in a DODAG represent another open problem. Briefly, RPL is a very encouraging routing protocol for LLNs. It gives a high flexibility level to consider various needs of underlying applications.

More research works are required to produce the best path in RPL by optimization. The operation mode is not explained in the drafts of IETF. The approaches that permit the node to choose an appropriate instance and to leave from one instance to another. It is important to specify the security methods like authentication and key management approaches provide a secure join to an instance in the network. The way by which the key is obtained by the node and key establishing and maintenance processes are not defined. Furthermore, the security configuration methods for the handling of the incoming packets represent an open issue.

8 Conclusion

The Internet of Things (IoT) will change the world and makes our environment completely monitored using smartly connected nodes. RPL Routing in IoTs networks has been extensively studied in the last years. This network is highly unreliable, prone to multihop interference, and to the quality of a time-varying link quality. Furthermore, the nodes are composed of limited memory, processing, power, and battery. In this article, the RPL routing standard with its principal characteristics. The DODAG formation is described. We have discussed the routing protocols for Low-Power and Lossy Networks. Several single paths and multipath routing protocols are reviewed. We have addressed the main challenges of RPL routing protocols in LLNs. Finally, we have suggested open research issues and future directions for the RPL routing in IoTs.

References

1. El Korbi, I., Brahim, M.B., Adjihy, C., Saidane, L.A.: Mobility enhanced RPL for wireless sensor networks. In: 2012 IEEE Third International Conference on the Network of the Future (NOF), Gammarth, Tunisie, 21–23 November 2012, pp. 1–8 (2012)
2. Lee, K.C., Sudhaakar, R., Dai, L., Addepalli, S., Gerla, M.: RPL under mobility. In: Proceedings of 2012 IEEE Consumer Communications and Networking Conference (CCNC), Las Vegas, NV, USA, January 2012, pp. 300–304 (2012). https://doi.org/10.1109/ccnc.2012.6181106
3. Tian, B., Hou, K.M., Shi, H., Liu, X., Diao, X.: Application of modied RPL under VANET-WSN communication architecture. In: Fifth (ICCIS 2013), pp. 1467–1470 (2013). https://doi.org/10.1109/iccis
4. Pavkovic, B., et al.: Efficient topology construction for RPL over IEEE 802.15. 4 in wireless sensor networks. Ad Hoc Netw. **15**, 25–38 (2014)
5. Le, Q., Ngo-Quynh, T., Magedanz, T.: RPL-based multipath routing protocols for Internet of Things on wireless sensor networks. In: 2014 International Conference on Advanced Technologies for Communications (ATC). IEEE (2014)
6. Marques, B.F., Ricardo, M.P.: Improving the energy efficiency of WSN by using application-layer topologies to constrain RPL-defined routing trees. In: 2014 13th Annual Mediterranean Ad Hoc Networking Workshop (MED-HOC-NET). IEEE (2014)
7. Jeong, G.K., Chang, M.: MoMoRo: providing mobility support for low-power wireless applications. IEEE Syst. J. **PP**(99), 1–10 (2014). ISSN: 1932-8184
8. Gaddour, O., Koubaa, A., et al.: Co-RPL: RPL routing for mobile low power wireless sensor networks using corona mechanism. In: 2014 9th IEEE International Symposium on Industrial Embedded Systems (SIES), pp. 200–209 (2014). https://doi.org/10.1109/sies.2014.6871205
9. Banh, M., et al.: Performance evaluation of multiple RPL routing tree instances for Internet of Things applications. In: 2015 International Conference on Advanced Technologies for Communications (ATC). IEEE (2015)
10. Djedjig, N., Tandjaoui, D., Medjek, F.: Trust-based RPL for the Internet of Things. In: 2015 IEEE Symposium on Computers and Communication (ISCC). IEEE (2015)
11. Iova, O., Theoleyre, F., Noel, T.: Exploiting multiple parents in RPL to improve both the network lifetime and its stability. In: 2015 IEEE International Conference on Communications (ICC). IEEE (2015)
12. Zhao, M., Ho, I.W.-H., Chong, P.H.J.: An energy-efficient region-based RPL routing protocol for low-power and lossy networks. IEEE Internet Things J. **3**(6), 1319–1333 (2016)
13. Aljarrah, E., Yassein, M.B., Aljawarneh, S.: Routing protocol of low-power and lossy network: survey and open issues. In: International Conference on Engineering and MIS (ICEMIS). IEEE (2016)
14. Lassouaoui, L., et al.: Evaluation of energy aware routing metrics for RPL. In: 2016 IEEE 12th International Conference on Wireless and Mobile Computing, Networking and Communications (WiMob). IEEE (2016)

15. Banh, M., et al.: Energy balancing RPL-based routing for Internet of Things. In: 2016 IEEE Sixth International Conference on Communications and Electronics (ICCE). IEEE (2016)
16. Qasem, M., et al.: A new efficient objective function for routing in Internet of Things paradigm. In: 2016 IEEE Conference on Standards for Communications and Networking (CSCN). IEEE (2016)
17. Jin, Y., et al.: Content centric routing in IoT networks and its integration in RPL. Comput. Commun. **89**, 87–104 (2016)
18. Oliveira, A., Vazão, T.: Low-power and lossy networks under mobility: a survey. Comput. Netw. **107**, 339–352 (2016)
19. Iova, O., et al.: RPL: the routing standard for the Internet of Things … or is it? IEEE Commun. Mag. **54**(12), 16–22 (2016)
20. Iova, O., et al.: The love-hate relationship between IEEE 802.15. 4 and RPL. IEEE Commun. Mag. **55**(1), 188–194 (2017)
21. Zhao, M., Chong, P.H.J., Chan, H.C.B.: An energy-efficient and cluster-parent based RPL with power-level refinement for low-power and lossy networks. Comput. Commun. **104**, 17–33 (2017)
22. Ghaleb, B., et al.: A new enhanced RPL based routing for Internet of Things. In: 2017 IEEE International Conference on Communications Workshops (ICC Workshops). IEEE (2017)
23. Kim, H.-S., et al.: Load balancing under heavy traffic in RPL routing protocol for low power and lossy networks. IEEE Trans. Mob. Comput. **16**(4), 964–979 (2017)
24. Bouaziz, M., Rachedi, A., Belghith, A.: EC-MRPL: an energy-efficient and mobility support routing protocol for Internet of Mobile Things. In: 2017 14th IEEE Annual Consumer Communications and Networking Conference (CCNC). IEEE (2017)
25. Fotouhi, H., Moreira, D., Alves, M.: mRPL: boosting mobility in the Internet of Things. J. Ad Hoc Netw. **26**, 17–35 (2015)
26. Zhao, M., et al.: A comprehensive study of RPL and P2P-RPL routing protocols: implementation, challenges and opportunities. Peer-to-Peer Netw. Appl. **10**(5), 1232–1256 (2017)
27. Gaddour, O., Koubâa, A., Abid, M.: Quality-of-service aware routing for static and mobile IPv6-based low-power and lossy sensor networks using RPL. Ad Hoc Netw. **33**, 233–256 (2015)
28. Mostafa, B., et al.: Distributed monitoring in 6LoWPAN based Internet of Things. In: 2016 International Conference on Selected Topics in Mobile and Wireless Networking (MoWNeT). IEEE (2016)
29. Winter, T.: RPL: IPv6 routing protocol for low-power and lossy networks (2012)
30. Tripathi, J.: On Design, Evaluation and Enhancement of IP-Based Routing Solutions for Low Power and Lossy Networks. Drexel University, Philadelphia (2014)
31. Gaddour, O., Koubâa, A.: RPL in a nutshell: a survey. Comput. Netw. **56**(14), 3163–3178 (2012)
32. Winter, T., Thubert, P., et al.: RPL: IPv6Routing Protocol for Low Power and Lossy Networks. IETF Request for Comments 6550, March 2012
33. Vasseur, J.-P., Dunkels, A.: Interconnecting Smart Objects with IP – The Next Internet. Morgan Kaufmann, Burlington (2010)
34. Thubert, E.P.: RPL Objective Function 0. IETF Internet Draft: draft-ietf-roll-of 0-03 (2010). http://tools.ietf.org/html/draft-ietf-roll-of0-12
35. Brachman, A.: RPL objective function impact on LLNs topology and performance. In: Balandin, S., Andreev, S., Koucheryavy, Y. (eds.) NEW2AN/ruSMART-2013. LNCS, vol. 8121, pp. 340–351. Springer, Heidelberg (2013). https://doi.org/10.1007/978-3-642-40316-3_30

36. Thubert, P.: Objective function zero for the routing protocol for low-power and lossy networks (RPL) (2012)
37. Conta, A., Deering, S., Gupta, E.M.: Internet Control Message Protocol (ICMPv6) for the Internet Protocol Version 6 (IPv6) Specification, RFC: 4443, March 2006
38. Martocci, E.J., De Mil, P., Riou, N., Vermeylen, W.: Building Automation Routing Requirements in Low-Power and Lossy Networks. RFC: 5867, March 2010
39. Brandt, A., Buron, J., Porcu, G.: Home Automation Routing Requirements in Low-Power and Lossy Networks. Request for Comments: 5826, March 2010
40. Pister, E.K., Thubert, E.P., Dwars, S., Phinney, T.: Industrial Routing Requirements in Low-Power and Lossy Networks, RFC: 5673, March 2009
41. Dohler, E.M., Watteyne, E.T., Winter, E.T, Barthel, E.D.: Routing Requirements for Urban Low-Power and Lossy Networks. Request for Comments: 5548, March 2009
42. Lu, Y.: Is IPv6 a better choice to handle mobility in wireless sensor networks? Dissertation (2013)
43. Fotouhi, H., Zuniga, M., Alves, M., Koubaa, A., Marrón, P.: Smart-HOP: a reliable handoff mechanism for mobile wireless sensor networks. In: Picco, G.P., Heinzelman, W. (eds.) EWSN 2012. LNCS, vol. 7158, pp. 131–146. Springer, Heidelberg (2012). https://doi.org/10.1007/978-3-642-28169-3_9
44. Stojmenovic, I., Lin, X.: Loop-free hybrid single-path/flooding routing algorithms with guaranteed delivery for wireless networks. IEEE Trans. Parallel Distrib. Syst. **12**, 1023–1032 (2001)
45. Lee, K.C., Sudhaakar, R., Ning, J., Dai, L., Addepalli, S., Gerla, M.: A comprehensive evaluation of RPL under mobility. Int. J. Veh. Technol., 300–304 (2012)
46. Tian, B., et al.: Application of modified RPL under VANET-WSN communication architecture. In: 2013 Fifth International Conference on Computational and Information Sciences (ICCIS). IEEE (2013)
47. Idrees, A.K., Deschinkel, K., Salomon, M., Couturier, R.: Multiround distributed lifetime coverage optimization protocol in wireless sensor networks. J. Supercomput. **74**, 1–24 (2017)
48. Idrees, A.K., Deschinkel, K., Salomon, M., Couturier, R.: Perimeter-based coverage optimization to improve lifetime in wireless sensor networks. Eng. Optim. **48**(11), 1951–1972 (2016)
49. Harb, H., Idrees, A.K., Jaber, A., Makhoul, A., Zahwe, O., Taam, M.A.: Wireless sensor networks: a big data source in Internet of Things. Int. J. Sens. Wirel. Commun. Control **7**(2), 93–109 (2017)
50. Idrees, A.K., Deschinkel, K., Salomon, M., Couturier, R.: Distributed lifetime coverage optimization protocol in wireless sensor networks. J. Supercomput. **71**(12), 4578–4593 (2015)
51. Idrees, A.K., Harb, H., Jaber, A., Zahwe, O., Taam, M.A.: Adaptive distributed energy-saving data gathering technique for wireless sensor networks. In: Wireless and Mobile Computing, Networking and Communications (WiMob), pp. 55–62. IEEE, October 2017
52. Idrees, A.K., Al-Qurabat, A.K.M.: Distributed adaptive data collection protocol for improving lifetime in periodic sensor networks. IAENG Int. J. Comput. Sci. **44**(3) (2017)
53. Adat, V., Gupta, B.B.: Security in Internet of Things: issues, challenges, taxonomy, and architecture. Telecommun. Syst. **67**(3), 423–441 (2018)
54. Pooranian, Z., Shojafar, M., Naranjo, P.G.V., Chiaraviglio, L., Conti, M.: A novel distributed fog-based networked architecture to preserve energy in fog data centers. In: 2017 IEEE 14th International Conference on Mobile Ad Hoc and Sensor Systems (MASS), pp. 604–609. IEEE, Florida (2017)
55. Ahmadi, A., Shojafar, M., Hajeforosh, S.F., Dehghan, M., Singhal, M.: An efficient routing algorithm to preserve k-coverage in wireless sensor networks. J. Supercomput. **68**(2), 599–623 (2014)

System and Network Security

A New RGB Image Encryption Based on DNA Encoding and Multi-chaotic Maps

Sarab M. Hameed$^{(\boxtimes)}$ ⓘ and Ibtisam A. Taqi ⓘ

Collage of Science, University of Baghdad, Baghdad, Iraq
sarab_majeed@yahoo.com, iat812000@yahoo.com

Abstract. This paper introduces a new diffusion and confusion scheme for encrypting an image by utilizing Deoxyribonucleic Acid (DNA) sequence and chaotic maps. First, the plain image is transformed into three components and each component is confused by DNA encoded to generate DNA matrices for a colour image. Next, the DNA colour image is permuted using a Chen hyper-chaotic map. In addition, a keystream is produced by a combination of new Beta and Sine chaotic maps. Conclusively, to obtain the cipher image, DNA XOR operation between the generated keystream and permuted DNA image is performed. The results indicate that the proposed scheme is secure against several attacks and provides large secret keyspace and very responsive to slight changes to the secret key.

Keywords: Beta map · Chen hyper-chaotic map · Sine map · DNA encoding

1 Introduction

With the exponential growth of multimedia and network technologies, digital image processing has been vastly utilized in all aspects of human life, such as remote sensing, industrial inspection, the medical field, meteorology, communications, surveillance, and intelligent robots. Consequently, increasing attention has been directed to image information. Additionally, it is essential for the protection of image data, especially in military, commercial, and medical fields. An efficient means to guard the transmission of digital images is using encryption technology. Image data has the characteristics of large amounts of data, strong correlations, and high redundancy. The existing classical encryption methods cannot meet the needs of image encryption because of their low efficiency and security [1].

The popularity of different kinds of image encryption methods with chaos maps is increased due to the chaos maps features related to confusion and diffusion properties of a good cipher. Also, DNA represents a new and promising direction in cryptography research. DNA can be employed in cryptography for storing and carrying the information, in addition to computation [2].

A number of efforts for image encryption have been suggested in the literature. Liu et al. in [3] used DNA complementary rule and chaotic maps for image encryption. MD5 hash initialized the chaotic maps. Furthermore, to diffuse the original image, a piecewise linear chaotic map (PWLCM) was used. Then the resulted diffused image was encoded DNA coding. The results show that the scheme provides large key space

© Springer Nature Switzerland AG 2018
S. O. Al-mamory et al. (Eds.): NTICT 2018, CCIS 938, pp. 69–85, 2018.
https://doi.org/10.1007/978-3-030-01653-1_5

to counter against general attacks. Liu et al. in [4] used 1D logistic map chaotic map and DNA encoding for encryption an RGB image. The proposed algorithm removes the correlation between pixels using DNA addition operation and disturbs the pixels value of a chaotic system. Experimental results demonstrate that the proposed algorithm is simple and effective to counter exhaustive attack and statistical attack. Zhang et al. in [5] suggested image fusion encryption algorithm using DNA sequence operations and 1D logistic, sine maps. Chen hyper-chaotic system is used to shuffle the positions of pixels. Testing the performance of the proposed algorithm was performed on grey images. The experimental results illustrate the ability of the proposed algorithm to counter statistical analysis and exhaustive attacks. Mokhtar et al. in [6] proposed coloured image encryption algorithm using DNA sequence as a one-time pad (OTP) for changing the values of pixel and 1D logistic map to confuse and diffuse image pixels. The results illustrate the ability of the proposed algorithm to withstand differential attack and statistical attack. Kumar et al. in [7] proposed a new algorithm that used a compound of chaotic maps with diffusion for image encryption. The results illustrate that the proposed algorithm is proper for image encryption with high security. Wang et al. in [8] used coupled map lattice (CML) and DNA sequence operations for encrypting an image. The DNA encoding was used to confuse the image pixels. The results show that the proposed method improves the security of the image. Niyat et al. in [9] suggested an encryption algorithm for RGB image using DNA sequence operation and chaotic maps, namely, Chen hyper-chaotic system for image shuffling, 1D logistic and sine maps for key generation. The results confirm the capability of the algorithm to counter several attacks. Wang et al. in [10] proposed an encryption scheme for colour image using 1D and 2D logistic map for generating a chaotic matrix. Then, the two chaotic maps are repeated one after the other for permuting the matrix. The experimental results prove the security and suitability of the scheme for image encryption. Zahmoul et al. in [11] proposed a Beta chaotic map that generates chaotic sequences for encryption. The image pixels position is shuffled with different pseudo-random sequences to obscure the correlation between the cipher and the original images. The proposed algorithm is capable of thwarting many attacks. Niyat et al. in [12] used hyper-chaotic system and cellular automata for colour image encryption. The results observe that the encrypted image produces a uniform histogram and small correlation between pixels and large key space. Furthermore, the algorithm can withstand different attacks including differential attacks, statistical analysis, comprehensive attacks, data lost attack and noise with different intensity.

The rest of the paper is organized as follows. In Sect. 2, preliminary concepts related to DNA encoding, decoding and chaotic maps are presented. Section 3 presents the suggested RGB image encryption and decryption scheme. Section 4 illustrates the evaluation of the results regarding quality of image and security. Section 5 provides a conclusion of the presented work and some scope for future work.

2 Preliminaries

2.1 Chaotic Maps

The suggested RGB image encryption scheme utilizes the chaotic maps namely, Beta chaotic, Sine chaotic and Chen hyper-chaotic maps.

The Beta chaotic map can be used in many applications, such as image compression, image object detection, and for various distinct applications. Beta map is defined as in Eq. (1) [11]

$$\text{Beta}(x, x_1, x_2, p, q) = \begin{cases} \left(\frac{x-x_1}{x_c-x_1}\right)^p \left(\frac{x_2-x}{x_2-x_c}\right)^q & \text{if } x \in [x_1, x_2] \\ 0 & \text{otherwise} \end{cases} \tag{1}$$

$$x_c = \frac{(px_2 + qx_1)}{(p+q)} \tag{2}$$

$$p = b_1 + c_1 \times a \tag{3}$$

$$q = b_2 + c_2 \times a \tag{4}$$

Where

$$b1 = 3, b2 = 5, c1 = 1, c2 = -1, a = 0.7$$

Sine map can is similar to logistic map and is described by Eq. (5) [9]:

$$y_{n+1} = r Sin(\pi y_n) \tag{5}$$

Where

$$r \in (0, 4], y_n \in (0, 1), n = 0, 1, 2, \ldots$$

Chen hyper-chaotic map is one of the hyper chaotic functions that have large space keys and is described as below [9]

$$\left. \begin{array}{l} x = a(y - x) \\ y = -xz + dx + cy + q \\ z = xy - bz \\ q = x + k \end{array} \right\} \tag{6}$$

Where $a, b, c, d, and\, k$ are parameters of the map. $a = 36, b = 3, c = 28, d = 16$ and $-0.7 \le k \le 0.7$, $x_0 = 0.3, y_0 = -0.4, z_0 = 1.2, q_0 = 1, k = 0.2$.

2.2   DNA Operation and Rules

DNA is a class of organic macromolecule and is composed of nucleotides that hold a separate base. Four vital nucleic acids are used to perform DNA sequence. These are Adenine (A), Cytosine (C), Guanine (G) and Thymine (T). A constantly joins with T, and C constantly joins with G. The overall numbers of possible combinations are twenty-four and only eight follow the complementary rule as presented in Table 1. DNA encoding involves representing each nucleotide by binary number following the complementary rule [1].

Table 1. The encoding rules for DNA sequences.

DNA sequence	Rule							
	1	2	3	4	5	6	7	8
00	A	A	C	G	C	G	T	T
01	C	G	A	A	T	T	C	G
10	G	C	T	T	A	A	G	C
11	T	T	G	C	G	C	A	A

The addition and subtraction of DNA are done over modulo 4. The addition and subtraction are given in Table 2 when rule one is utilized [1]. The DNA-XOR operation is given in Table 3 [1]:

Table 2. DNA addition and subtraction operations

+	A	C	G	T	−	A	C	G	T
A	A	C	G	T	A	A	T	G	C
C	C	G	T	A	C	C	A	T	G
G	G	T	A	C	G	G	C	A	T
T	T	A	C	G	T	T	G	C	A

Table 3. DNA XOR operation

⊕	A	C	G	T
A	A	C	G	T
C	C	A	T	G
G	G	T	A	C
T	T	G	C	A

3   Suggested Image Encryption and Decryption

The proposed image encryption scheme as shown in Fig. 1 is based on the combining DNA sequence and chaotic maps. The proposed scheme consists of five processes.

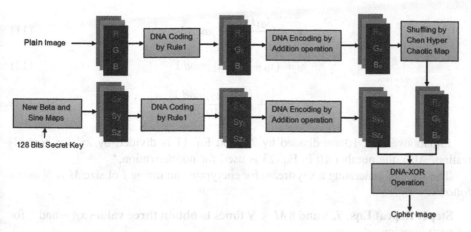

Fig. 1. General layout of the proposed scheme

3.1 Keystream Generation

A new method is suggested to generate a keystream to encrypt an image I. Secret key of 128-bit $K = \{k_1, k_2, \ldots k_{32}\}$ where k_i represents a 4-bit hexadecimal digit is used to generate the initial values of the chaotic maps.

A new Beta chaotic map inspired from the beta chaotic map is suggested as formulated in Eq. 7:

$$\text{newBeta}(x, x_1, x_2, p, q) = k \times \left(\left(\frac{x - x_1}{x_c - x_1} \right)^p \left(\frac{x_2 - x}{x_2 - x_c} \right)^q \right) mod\, x_2 \qquad (7)$$

Where $k = 0.99$

The suggested new Beta chaotic map in Eq. 7, Sine chaotic map in Eq. 5 and a new chaotic map that combines new Beta and Sine chaotic maps as formulated in 8 are formed for sequence generation

$$z_n = \left(\text{newBeta}(x_n, x_1, x_2, p, q) + \frac{(4 - r)Sin(\pi y_n)}{4} \right) mod\, 1 \qquad (8)$$

The initial value for new Beta chaotic maps is considered from a secret key using the suggested equations in 9, 10 and 11. Moreover, formulas 12 and 13 are suggested to initialize Sine chaotic maps.

$$x_0 = \left(\frac{k_1 k_2 \ldots k_{10}}{2^{23}} \right) mod\, 1 \qquad (9)$$

$$x_1 = \left(\frac{k_{11} k_{12} \ldots k_{20}}{2^{23}} \right) mod\, 1 \qquad (10)$$

$$x_2 = \left(\frac{k_{21}k_{22}..k_{32}}{2^{24}}\right) mod\ 1 \tag{11}$$

$$y_0 = (x_0 + x_1 + x_2) mod\ 1 \tag{12}$$

$$r = 3.9 + \frac{(x_1 + x_2) mod\ 1}{10} \tag{13}$$

Equations 9 and 10 are divided by 2^{23} and Eq. 11 is divided by 2^{24} for normalization. Also, the number 10 in Eq. 13 is used for normalization.

The steps of generating a keystream for encrypting an image I of size $M \times N$ are as follows:

Step1: Repeat Eqs. 7, 5 and 8 $M \times N$ **times to obtain three values** x_i, y_i and z_i for each iteration.

Step2: Generate three sequences $S_x = \{x_1, x_2, \ldots x_{M \times N}\}, S_y = \{y_1, y_2, \ldots y_{M \times N}\}$ and $S_z = \{z_1, z_2, \ldots z_{M \times N}\}$ **of length** $M \times N$ **and in** the range $[0, 255]$ according to Eqs. 14, 15 and 16 respectively.

$$x_i = \lfloor (|x_i| \times 10^{14}) mod\ 256 \rfloor \tag{14}$$

$$y_i = \lfloor (|y_i| \times 10^{14}) mod\ 256 \rfloor \tag{15}$$

$$z_i = \lfloor (|z_i| \times 10^{14}) mod\ 256) \rfloor \tag{16}$$

Step3: Transform the sequences S_x, S_y and S_z into binary representation. After that, the binary numbers of are converted to DNA sequence using Rule 1 in Table 1 to obtain three DNA matrices $S_{xc}(M \times 4N), S_{yc}(M \times 4N)$ and $S_{zc}(M \times 4N)$ that are expressed in four nucleic acid bases.

Step4: Perform DNA addition to get three DNA encoded matrices S_{xe}, S_{ye} and S_{ze} as follows:

$$\left. \begin{array}{l} S_{xe} = S_{xc} + S_{yc} \\ S_{ye} = S_{yc} + S_{zc} \\ S_{ze} = S_{zc} + S_{ye} \end{array} \right\} \tag{17}$$

3.2   Confusion by DNA Addition Operation

Confusion property is one characteristic that should be performed for image encryption. Accordingly, DNA encoding is utilized to achieve this property by the following steps:

Step1: The plain image I **is disintegrated into three** $R(M, N), G(M, N)$ **and** $B(M, N)$ **components, as shown below:**

$$R = \{r_1, r_2, \ldots r_{M \times N}\}$$
$$G = \{g_1, g_2, \ldots g_{M \times N}\}$$
$$B = \{b_1, b_2, \ldots b_{M \times N}\}$$

Where r_i, g_i and b_i are the ith pixel values in the range $[0, 225]$ for red, green and blue components

Step2: Convert the disintegrated matrices of R, G, B from decimal form into binary matrices $R_b(M, N \times 8)$, $G_b(M, N \times 8)$ and $B_b(M, N \times 8)$

Step3: Apply DNA coding using rule 1 in Table 1 to produce three DNA coding matrices $R_c.(M, N \times 4)$, $G_c(M, N \times 4)$ and $B_c(M, N \times 4)$ and then encoding them by DNA addition operation to yield three DNA encoding matrices $R_e(M, N \times 4)$, $G_e(M, N \times 4)$ and $B_e(M, N \times 4)$ as follows:

$$\left. \begin{array}{l} R_e = R_c + G_c \\ G_e = G_c + B_c \\ B_e = B_c + G_e \end{array} \right\} \tag{18}$$

3.3   Image Diffusion

Diffusion shows an essential role in image encryption. The diffusion property for an image can be gained by destroying the correlation among the neighbouring pixels. Therefore, Chen hyper-chaotic map as in [5] that uses four-order Rung–Kutta method to get x, y, z, q sequences and then the decimal part of the element is preserved and the integer part is excluded to shuffle each pixel of an image.

$$\left. \begin{array}{l} [x_{ind}, x'] = sort(x) \\ [y_{ind}, y'] = sort(y) \\ [z_{ind}, z'] = sort(z) \\ [q_{ind}, q'] = sort(q) \end{array} \right\} \tag{19}$$

Where x', y', z' and q' is the new sequence afterward ascending x, y, z, and q respectively to verify the generation of all indexes to avoid data loss, $x_{ind}, y_{ind}, z_{ind}$, and q_{ind} are the index value of x', y', z' and q' respectively.

$$\left. \begin{array}{l} \forall i, 1 \leq i \leq M \text{ and } \forall j, 1 \leq j \leq 4N \text{ and} \\ R_S(i,j) = R_e(x_{ind}(i), y_{ind}(j)) \\ G_S(i,j) = G_e(x_{ind}(i), z_{ind}(j)) \\ B_S(i,j) = B_e(z_{ind}(i), q_{ind}(j)) \end{array} \right\} \tag{20}$$

3.4   Confusion by DNA XOR Operation

Finally, extra confusion is applied to produce a high level of security and make the pixel sensitive to the key. The confusion is applied by performing DNA XOR operation between the generated sequences S_{xe}, S_{ye} and S_{ze} and R_s, G_s, B_s to an encrypted image C as follows:

$$\left.\begin{array}{l} C_R = R_s \oplus S_{xe} \\ C_G = G_s \oplus S_{ye} \\ C_B = B_s \oplus S_{ze} \end{array}\right\} \tag{21}$$

3.5   Image Decryption

Image decryption process is similar to the encryption process but the steps are taken by the reverse order as follows:

Step 1: Perform DNA XOR operation between C_R, C_G and C_B and the corresponding S_{xe}, S_{ye}, S_{ze} to obtain the shuffling matrices R_s, G_s and B_s as follows:

$$\left.\begin{array}{l} R_s = C_R \oplus S_{xe} \\ G_s = C_G \oplus S_{ye} \\ B_s = C_B \oplus S_{ze} \end{array}\right\} \tag{22}$$

Step 2: Perform inverse shuffling for R_s, G_s and B_s to get DNA encoding matrices R_e, G_e and B_e as follows:

$$\begin{array}{l} \forall i, 1 \leq i \leq M \text{ and } \forall j, 1 \leq j \leq 4N \text{ and} \\ \left.\begin{array}{l} R_e(x_{ind}(i), y_{ind}(j)) = R_S(i,j) \\ G_e(x_{ind}(i), z_{ind}(j)) = G_S(i,j) \\ B_e(z_{ind}(i), q_{ind}(j)) = B_S(i,j) \end{array}\right\} \end{array} \tag{23}$$

Step 3: Apply DNA Subtraction operation to get DNA coding matrices R_c, G_c and B_c as follows:

$$\left.\begin{array}{l} B_c = B_e - G_e \\ G_c = G_e - B_c \\ R_c = R_e - G_c \end{array}\right\} \tag{24}$$

Step 4: Convert the three matrices obtained from step 3 to binary then to decimal to get the plain image.

4   Experimental Results

The proposed scheme is coded in Visual C Sharp and the experiments are conducted on a *Lenovo Laptop* with Intel(R) Core (TM) i7-5500U, CPU @ 2.40 GHz and a Memory of 16.0 GB RAM and 64-bit system type. The performance of the proposed method is evaluated using the image dataset of the Signal and Image Processing Institute (SIPI) maintained by the University of South California (USC) [13]. Figure 1 depicts the Lena

(512×512), Baboon (256×256), and Pepper (256×256) and their corresponding cipher images. The results show that the cipher images are noisy and any information cannot be acquired, which means that the proposed scheme provides a good encryption result (Fig. 2).

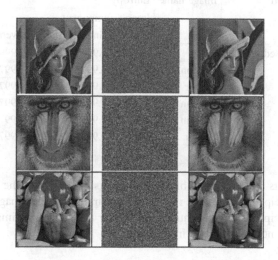

Fig. 2. Images encryption and decryption with the proposed scheme

4.1 Key Space Analysis

The key space size describes the ability to counter brute-force attack. The key space size for the proposed scheme is 2^{128}. Additionally, x, y, z and q parameters of Chen hyper-chaotic map, x_0, x_1, x_2, k of new Beta map and y and r of Sine map are considered as a secret key with the precision 10^{-14}. Therefore, the key space is $((10^{(14*3)} = 2^{140}) * 2^{24} * 2^{128} = 2^{292})$ that appears to be sufficient to counter brute-force attacks.

4.2 Statistical Attack

The proposed encryption scheme is evaluated regarding entropy, histogram analysis, and correlation coefficient to illustrate the ability of the proposed scheme to resist statistical attack.

Information entropy (H) measures the randomness of an image as expressed in Eq. 25.

$$H(m) = -\sum_{i=0}^{L} p(m_i) log_2 p(m_i) \tag{25}$$

Where $L = 255, m_i$ is i^{th} pixel value of an image, and $p(m_i)$ is m_i probability.

As clarified in Table 4, the entropy of the encrypted image using the proposed scheme is closer to 8 and slightly larger than the schemes in [9] and [12] which indicates the randomness of the cipher image is very high and the probability of deducing any information is too small. This means the proposed scheme provides a good encryption and can resist the statistical attack.

Table 4. The entropy of cipher images of the proposed scheme against [9] and [12]

Method	Image name	Entropy			
		Image component			
		Red	Green	Blue	Average
Proposed scheme	Lena	7.9992	7.9993	7.9993	7.9993
	Baboon	7.9972	7.9973	7.9974	7.9973
	Pepper	7.9972	7.9975	7.9972	7.9973
[9]	Baboon	7.9973	7.9968	7.9976	7.9972
[12]	Baboon	7.9972	7.9972	7.9972	7.9972
	Pepper	7.9971	7.9975	7.9974	7.9973

Figure 3 depicts red, green and blue components histogram of the Lena image and its corresponding cipher image. The results demonstrate the Lena image histogram and its corresponding cipher image is entirely dissimilar and the cipher image histogram is very flat and does not provide any information.

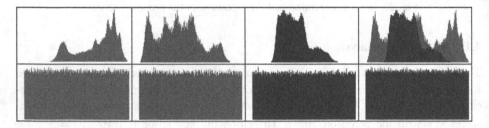

Fig. 3. Red, green and blue components histogram of the Lena image and its corresponding cipher image. (Color figure online)

Images are very correlated with their neighbouring pixel whether in horizontal, vertical and diagonal direction. Correlation coefficient (CC) measures the correlation of adjacent pixels and is calculated as follows [9]:

$$r_{xy} = \frac{cov(x,y)}{\sqrt{D(x)D(y)}} \tag{26}$$

$$E(x) = \frac{1}{N}\sum_{i=1}^{N} x_i$$

$$D(x) = \frac{1}{N} \sum_{i=1}^{N} (x_i - E(x_i))^2$$

$$cov(x, y) = \frac{1}{N} \sum_{i=1}^{N} (x_i - E(x_i))(y_i - E(y_i))$$

Where x and y are two neighbouring pixels values in the image, $D(x)$ is the variance, and N is the number of chosen neighbouring pixels of the image.

For testing purpose, 1000 pairs of neighbouring pixels are randomly selected in three directions from the original image and the corresponding cipher image. Figures 4 and 5 depict the correlation coefficients of each direction of Lena and the corresponding cipher image respectively. The figures illustrate the plain image correlation distribution is aggregated alongside the diagonal. However, cipher image correlation distribution is spread over the plane. Tables 5, 6 and 7 report the cipher image correlated coefficient in horizontal, vertical and diagonal directions. The result of correlation coefficients is so small in the proposed scheme when compared with the schemes in [7, 9] and [12], which indicates that the correlation is significantly reduced in the ciphered image and

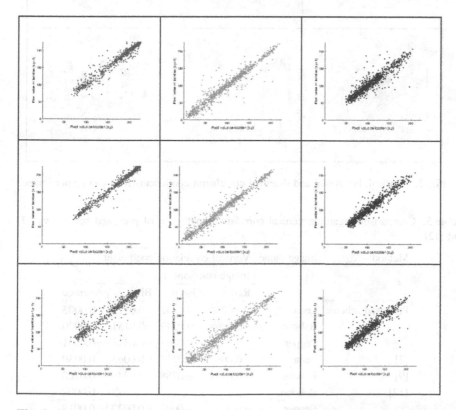

Fig. 4. Vertical, horizontal and diagonal correlation coefficients for Lena plain image.

the proposed scheme is capable of resisting the statistical attacks better than other schemes in [7, 9] and [12].

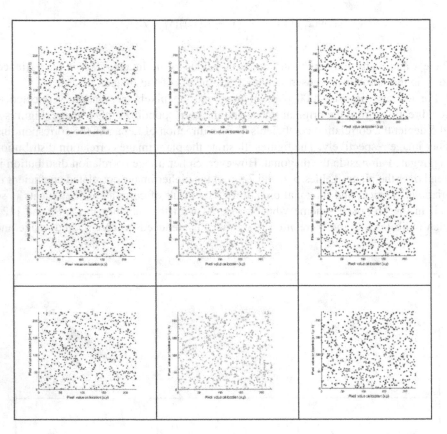

Fig. 5. Vertical, horizontal and diagonal correlation coefficients for Lena cipher image

Table 5. Comparison regarding vertical correlation coefficient of proposed scheme vs. [7, 9] and [12]

Method	Image name	Vertical correlation coefficient			
		Image component			
		Red	Green	Blue	Average
Proposed scheme	Lena	0.0230	−0.0342	0.0122	0.0003
	Baboon	−0.0098	0.0249	−0.0151	0.0000
	Pepper	0.0311	0.0003	−0.0318	−0.0001
[7]	Lena	−0.0099	0.0126	0.0063	0.0030
[9]	Baboon	0.0744	0.0788	0.0748	0.0760
[12]	Baboon	–	–	–	0.0002
	Pepper	0.0031	0.0001	0.0022	0.0018

Table 6. Comparison regarding horizontal correlation coefficient of proposed scheme vs. [7, 9] and [12]

Method	Image name	Horizontal correlation coefficient			
		Image component			
		Red	Green	Blue	Average
Proposed scheme	Lena	−0.0380	−0.0019	0.0379	−0.0007
	Baboon	−0.0459	0.0017	0.0433	−0.0003
	Pepper	−0.0094	0.0075	−0.0087	−0.0035
[7]	Lena	0.0181	−0.0067	0.0154	0.0089
[9]	Baboon	0.0761	0.0827	0.0757	0.0782
[12]	Baboon	–	–	–	−0.0008
	Pepper	0.0049	0.0054	0.0053	0.0052

Table 7. Comparison regarding diagonal correlation coefficient of proposed scheme vs. [7, 9] and [12]

Method	Image name	Diagonal correlation coefficient			
		Image component			
		Red	Green	Blue	Average
Proposed scheme	Lena	0.0313	−0.0158	−0.0159	−0.0001
	Baboon	0.0018	−0.0429	0.0395	−0.0005
	Pepper	0.0007	0.0416	−0.0401	0.0007
[7]	Lena	0.0085	0.0127	−0.0155	0.0019
[9]	Baboon	0.0733	0.0687	0.0701	0.0707
[12]	Baboon	–	–	–	−0.0006
	Pepper	0.0007	0.0017	0.0007	0.0010

Table 8. Comparison regarding UACI of the proposed scheme against [7] and [12]

Method	Image name	UACI			
		Image component			
		Red	Green	Blue	Average
Proposed scheme	Lena	33.4470	33.5013	33.3817	33.4433
	Baboon	33.5823	33.5536	33.3254	33.4871
	Pepper	33.4693	33.5873	33.2650	33.4405
[7]	Lena	33.2829	33.3459	33.3270	33.3186
[12]	Baboon	33.4753	33.5090	33.4176	33.4673
	Pepper	33.4570	33.4705	33.4423	33.4566

4.3   Differential Attack

The sensitivity of a cipher image should be high to the slight modifications in a plain image. To examine the impact of changing one pixel in the plain image on the cipher image, Number of Pixels Change Rate (NPCR) and Unified Average Changing Intensity (UACI) metrics are used. NPCR and UACI are calculated as in Eqs. 27 and 28 respectively.

$$UACI(C1, C2) = \frac{\sum_{i=1}^{M} \sum_{j=1}^{N} |C1(i,j) - C2(i,j)|/255}{M \times N} \times 100 \qquad (27)$$

Where M and N are image width and height respectively and $C1$ and $C2$ are encrypted images of the plain image and the modified one.

$$NCPR(C1, C2) = \frac{\sum_{i=1}^{M} \sum_{j=1}^{N} D(i,j)}{M \times N} \times 100 \qquad (28)$$

Where

$$D(i,j) = \begin{cases} 0, if\ C1(i,j) = C2(i,j) \\ 1, if\ C1(i,j) \neq C2(i,j) \end{cases}$$

Tables 8 and 9 report UACI and NPCR of the proposed scheme against schemes in [7] and [12] respectively. Obviously, the results show NPCR of the proposed scheme is more near to 100% than [7] and [12], which means the sensitivity of the proposed scheme to the modification of plain image is high and it is effective to counter plaintext attack. Moreover, UACI is more close to 33.48 than [7] and [12], which means the capability of withstanding the differential attack of the proposed scheme is better than [7] and [12].

Table 9.   Comparison regarding NPCR of the proposed scheme against [7] and [12]

Method	Image name	NPCR			
		Image component			
		Red	Green	Blue	Average
Proposed scheme	Lena	99.6254	99.6086	99.6288	99.6209
	Baboon	99.6292	99.6292	99.6613	99.6399
	Pepper	99.6292	99.6292	99.6613	99.6399
[7]	Lena	99.5659	99.5658	99.5959	99.5759
[12]	Baboon	99.6536	99.6078	99.6520	99.6378
	Pepper	99.6357	99.6158	99.6247	99.6254

4.4   Other Analysis

The Mean Square Error (MSE) and Peak Signal-To-Noise Ratio (PSNR) metrics are adopted for evaluating the performance of the proposed encryption scheme. The MSE is a quantitative measure that clarifies the distinction between plain image and cipher image as in Eq. 29 [6].

$$MSE = \frac{1}{MN} \sum_{i=0,j=0}^{M,N} (P(i,j) - C(i,j))^2 \tag{29}$$

Where $P(i,j)$ is plain image pixel value, $C(i,j)$ is cipher image pixel value, M and N are the dimensions of the image.

The mathematical representation of the PSNR is as in Eq. 30 [14]

$$PSNR = 20 \log_{10} \left(\frac{255}{\sqrt{MSE}} \right) \tag{30}$$

Tables 10 and 11 report the *MSE* value and *PSNR* value of encrypted images and decrypted images using the proposed method. The results point out that *MSE* value between the cipher image and the plain image is large. Furthermore, the *MSE* value between the decrypted image and the plain image is zero. This means that the proposed encryption is lossless image encryption (i.e., the plain image can be obtained without any data loss). In addition, the results show that *PSNR* values between the original and encrypted images using the proposed method are small.

Table 10. MSE of the proposed scheme

Image name	MSE			
	Image component			
	Red	Green	Blue	Average
Lena	10633.1718	9058.0282	7155.6620	8948.9540
Baboon	8712.1856	7768.4930	9463.7375	8648.1387
Pepper	8010.1788	11297.3192	11111.0343	10139.5108

Table 11. PNSR of the proposed scheme

Image name	PSNR			
	Image component			
	Red	Green	Blue	Average
Lena	7.8642	8.5605	9.5843	8.6697
Baboon	8.7295	9.2274	8.3702	8.7757
Pepper	9.0944	7.6010	7.6733	8.1229

5   Conclusions

In this paper, a new encryption scheme for RGB image is proposed. The key of the proposed scheme is generated by utilizing the combination of sine chaotic map and proposed new Beta chaotic map. The generated key is encoded using the DNA encoding rule, then DNA XOR operation is applied with a shuffled confused image by Chen hyper-chaotic map to produce cipher image. Experimental results reveal that the proposed scheme is capable of countering various attacks including brute-force attack, statistical attack, differential attack, and plaintext attack. Furthermore, the proposed scheme provides large keyspace and is very sensitive to any simple change in the secret key. The process of integrating chaotic map with DNA encoding in the proposed method for colour image encryption provides a high-security compared with other methods in accordance with standard evaluation metrics such as entropy, UACI, NPCR and Correlation Coefficient. The results of the number of pixel change rate for all test images of the same size are stable which is difficult for an attacker to determine the encrypted image. Furthermore, the quality of the colour image is preserved without any loss of data as indicated by the MSE between the original and decrypted image. In addition, using DNA encoding for achieving confusion property and using the proposed chaotic map to produce diffusion property give robust and complex encryption method.

References

1. Zhang, X., Han, F., Niu, Y.: Chaotic image encryption algorithm based on bit permutation and dynamic DNA encoding. Comput. Intell. Neurosci. 11 (2017)
2. Wang, X.-Y., Zhang, Y.-Q., Zhao, Y.-Y.: A novel image encryption scheme based on 2-D logistic map and DNA sequence operations. Nonlinear Dyn. **82**, 1269–1280 (2015)
3. Liu, H., Wanga, X., Kadirc, A.: Image encryption using DNA complementary rule and chaotic maps. Appl. Soft Comput. **12**, 1457–1466 (2012)
4. Liu, L., Zhang, Q., Wei, X.: A RGB image encryption algorithm based on DNA encoding and chaos map. Comput. Electr. Eng. **38**, 1240–1248 (2012)
5. Zhang, Q., Guo, L., Wei, X.: A novel image fusion encryption algorithm based on DNA sequence operation and hyper-chaotic system. Optik **124**, 3596–3600 (2013)
6. Mokhtar, M.A., Gobran, S.N., El-Badawy, E.S.A-M.: Colored image encryption algorithm using DNA code and chaos theory. In: 5th International Conference on Computer & Communication Engineering (2014)
7. Kumar, M., Powduri, P., Reddy, A.: An RGB image encryption using diffusion process associated with chaotic map. J. Inf. Secur. Appl. **21**, 20–30 (2015)
8. Wang, X.-Y., Zhang, Y.-Q., Bao, X.-M.: A novel chaotic image encryption scheme using DNA sequence operations. Opt. Lasers Eng. **73**, 53–61 (2015)
9. Niyat, A.Y., HeiHei, R.M., Vafaei Jahan, M.: A RGB image encryption algorithm based on DNA sequence operation and hyper-chaotic system. In: International Congress on Technology, Communication and Knowledge (ICTCK) (2015)
10. Wang, X.-Y., Zhao, Y., Zhang, H., Guo, K.: A novel color image encryption scheme using alternate chaotic mapping structure. Opt. Lasers Eng. **82**, 79–86 (2016)

11. Zahmoul, R., Ejbali, R., Zaied, M.: Image encryption based on new Beta chaotic maps. Opt. Lasers Eng. **96**, 39–49 (2017)
12. Niyat, A.Y., Moattar, M.H., Torshiz, M.N.: Color image encryption based on hybrid hyper-chaotic system and cellular automata. Opt. Lasers Eng. **90**, 225–237 (2017)
13. Signal and Image Processing Insititute. University of South California. http://sipi.usc.edu/database/database.php?volume=misc. Accessed 1 Oct 2017
14. Kumar, M., Iqbal, A., Kumar, P.: A new RGB image encryption algorithm based on DNA encoding and elliptic curve Diffie-Hellman cryptography. Sig. Process. **125**, 187–202 (2016)

An Efficient E-ticket Fare
Scheme for Passengers Based on the Distance
Traveled Between Entry Point and Exit Point

Sattar J. Aboud[1(✉)] and Zinah S. Jabbar[2]

[1] University of Information Technology and Communications, Baghdad, Iraq
sattar_aboud@yahoo.com
[2] University of Imam Ja'afar Al-Sadiq, Baghdad, Germany

Abstract. This scheme determines the ticket price a passenger should pay based on the amount of distance traveled in each trip. The recent technology can assist to decrease the price of e-tickets as well as to enhance a control of the passenger movements. The proposed scheme provides robust privacy to non-fraud passengers, meaning that the authorized service provider is unable to reveal *id* of the passengers. Furthermore, various trips cannot be linked to the same passenger among themselves. But, such schemes should be protected from cheating and also protect passenger privacy. It means that the authorized service provider cannot be revealing *id* of passengers. But, the pseudonym of the passenger can be revoked when he cheats. The proposed scheme has been implemented and its results have been rated. The initial outcomes given in this paper appear very promising. The outcomes show that the scheme is applicable and will give good results in case of application to new generations of mobile phones. Also, the results show that the method is more efficient and faster than the already existing schemes.

Keywords: E-ticket · Pseudonym · Mobile applications · Distance-based
Public-key encryption · Privacy

1 Introduction

The efficient e-ticket scheme is proposed for high-intensity general transport passengers. The calculation of a charge by the points of entry and exit or distance-based scheme have several advantages such as excluding cash selling machines, providing comfort for passengers, speeding up passengers' transport, and decreasing the accounting costs of the public transport company. To do so, an implementation for an efficient scheme has become essential. Therefore, the trustworthy management of entry and exit of passengers in this scheme is important because passengers pay the transportation fees under such usage.

When a scheme can recognize the identity of passengers and determine their entry and exit points in the stations and whenever they leave and when they return, this means that a scheme is able to track their moves. Using such moves, it could establish a profile for passengers. That is a critical threat to privacy, in the sense that it breaches privacy of passengers. For this reason, the proposed scheme should maintain passenger

© Springer Nature Switzerland AG 2018
S. O. Al-mamory et al. (Eds.): NTICT 2018, CCIS 938, pp. 86–109, 2018.
https://doi.org/10.1007/978-3-030-01653-1_6

privacy by blocking traceability, recording and analysis of passenger characteristics. If the scheme is based solely on the availability of non-revocable privacy, in such a case some persons fleeing justice could use public transport to avoid prosecution by the security forces. Therefore, there are two conflicting demands. To find a compromise formula to meet the two demands is to authorize a scheme to revoke the pseudonym of the passenger identity after obtaining the court's approval. Another matter that needs to be considered in the proposed scheme is the slow pace of current schemes during entry and exit processes because of the large number of passengers using them. It is therefore necessary to accelerate this in the proposed scheme.

With respect to the appliances to be utilized in the proposed scheme, as long as the newest directions are moving towards the utilization of smartphones, rather than smart cards [2, 4]. So, it should be said that a Smartphone is the device required to be carried by the traveler when traveling.

The main aim of this project is to provide safe management of the proposed scheme, besides robust privacy to loyal passengers, that is, a service provider is unable to reveal *id* of his passengers and that various trips for the same passenger cannot link together. But, the identity of passengers may be revoked if they behave badly.

The aim also, is cash removal, excluding selling machines, providing greater comfort for passengers, quicker travel without delay and lessened accounting costs. Then, as a result, the scheme will make costs drop as well as better infrastructures control; this will be subject to passenger flows. But, when passengers act in a deceptive and unfair way their anonymity could be revealed for acquiring proper legal procedures against them if a scheme provides revocable anonymity. In addition, it will not require the passenger to get fresh credentials each time they would like to connect with the scheme. Such property enhances scheme usability, while former schemes require fresh credential evidence in each new trip.

2   Related Works

The author has carried out analysis of the existing schemes, which provide revocable anonymity to passengers [1–5]. The information analyzed is based on anonymity degree of passengers to ensure that privacy is not violated, the ability of schemes tracking different trips for the same passenger and the amount of information being disclosed about passengers as a result of trip tracking by the scheme, and ultimately the appliances utilized in existing schemes.

In most schemes analyzed, the authorized service provider is able to link various trips for the same passenger. With the linkable systems, a revelation *id* of a passenger in any trip causes a revelation of all trips for the same passenger. Actually, this indicates to a weak anonymity in these schemes. Therefore, the authorized service provider has knowledge of the direction of the passenger trips, whenever they will leave and when they return. Knowing such information about the passengers gives the authorized service provider the opportunity to generate profiles for them. Such profiles are beneficial for the authorized service provider, since they can use them to evolve the overall transport scheme. But, establishing profiles to passengers is breach of privacy, so the proposed scheme should avert follow up passengers and their privacy.

In 2006, Heydt-Benjamin *et al.*, [1] introduced a scheme called privacy for public transportation. In their scheme, the authorized service provider is unable to track the trips, and also the scheme requires a credential of fresh data for each trip. This means there are some additional costs on transportation that will be added on the passenger ticket. Another issue about credential regeneration is that it needs more complicated constructing. This means adding another additional cost in order that it can administer the large numbers of credentials.

Moreover, in 2008 Madlmayr *et al.*, [2] introduced an e-ticket scheme using proximity technology based on the security side of the system and the ticket distribution. This scheme is based on the browser plug-in combination with contactless radio frequency identification reader. Additionally, in 2013 Isern-Deyà *et al.*, [3] introduced a scheme that provides privacy for honest passengers. In their scheme, the authorized service provider cannot reveal *id* of its passengers and cannot link different trips for the same passenger. The scheme has been designed for use on personal mobile phones. In 2014 Arfaoui *et al.* [4] introduced another scheme, the privacy-preserving contactless mobile service, in which the passenger *id* cannot be linked to his actions when he uses a transport scheme. The security of the scheme is based on the merging of secure items in Smartphones with the privacy-enhancing cryptography protocol based on the user group of signatures. Also, in 2017 Sattar Aboud *et al.* [5] presented two schemes of e-ticket with anonymity in transferability also based-on the use of group signatures to show different degree of complexity.

The proposed scheme enables revocable anonymity and at the same time non-traceability of passengers. In addition, the scheme is structured to get benefit from the utilization of the private passengers' mobile phones. Furthermore, no passenger is required to get fresh credentials each time he takes a trip, as required in [1] because credential regeneration needs an additional charge.

3 Notations Used

The notations used are as follows

G_1, G_2 :	Bilinear groups
a_1, a_2 :	Two generators
$G_1 \times G_2 \rightarrow G_T$:	Bilinear map
H :	Secure hash function
$l, z, b \in Z_{G_1}$	Public values
P :	Passenger
$d \in G_2$:	Public value
$o \in Z_n$:	Integer value
$f_1, f_2 \in Z_n^*$:	Private values
(w_1, y_1) :	Strong Diffie-Hellman problem
KG :	Key generation
SG :	Signature generation
SV :	Signature verification
TS :	Tracing signature
LS :	Linkable signatures

VL :	Verification linkable
pk :	Public key
$sk[i] = (w_i, y_i)$:	Secret key
$m \in \{0,1\}^*$:	Message
id :	Passenger identity
ψ :	The signature
es :	Departure station of issuer id
ds :	Destination station of issuer id
TA_1 :	Trusted authority generates passenger and issuer accounts
TA_2 :	Trusted authority controls a revocation list and parameter keys
p, q :	Prime numbers public parameters
n :	Composite modulus
$\theta(n)$:	Theta = $(p-1)(q-1)$
e :	Public key for encryption
rl :	Revocation list
e_p :	Passenger anonymity name for payment
d_p :	Arbitrary integer private
$D(m)$:	Digital signature for a message m
D_p :	Digital signature of passenger
D_{TA_1} :	Digital signature of trusted authority TA_1
D_{TA_2} :	Digital signature of trusted authority TA_2
e_{in} :	Entry ticket
e_{out} :	Exit ticket
t_1 :	Departure timestamp
t_2 :	Exit timestamp
sb :	Serial number make by ds
e'_{in} :	Signature of entry ticket, signed by es
e'_{out} :	Signature of exit ticket
c :	Verification key
pk_{ds} :	Issuer public key in destination station ds
f :	Charges fee due
s_p :	Key for encryption of e_P
b :	The challenge
b' :	Signature of the challenge
y_5:	Zero knowledge proof of identity
pr :	Payment refusal
pr' :	Signature of payment refusal signed by TA_1
pa :	Payment acceptance
pa' :	Signature of payment acceptance

4 The Preliminaries

These days the schemes that depend on distance are widely used compared with the schemes based on time. An illustration of schemes based on the distance is the large-scale public transfer services like the underground and bus metro. With the schemes that rely on distance, a fee is determined based on a distance between entry and leaving stations. Such stations are related to the authorized service provider. It can use an algorithm that computes the charges due by adding another input element such as a type of coach, in hour or on day. Such schemes are more complicated compared with those calculated by time. Over time, it can secure the distance among two stations in both tracks.

As stated by an expert's report about some towns, the application of the e-ticket scheme in transportation is a challenge. With the advent of modern technology, like Smartphones, it can aid in offering appropriate schemes over the Internet. In section two, it is noticed that one of the disadvantages of already existing cost schemes based on distance is that the passengers cannot transform a direction of a motion without getting out of order. It means that they should leave in accordance with their ways. This will need a separate exit scheme from the track.

The proposed scheme is designed in a manner that does not require a separate exit scheme from direction. The scheme utilizes identity-based multi-signature from RSA [6] scheme to check that the passenger is a real partner from the selected passengers group. The identity-based multi-signature from RSA is illustrated in Sect. 5, with the proposed manner which lets the pair of group signatures to be linked. Then, the proposed distance-based scheme is described in Sect. 6. Section 7 illustrates an analysis for the proposed scheme. Lastly, a conclusion is given in Sect. 8.

The proposed scheme uses bilinear groups G_1, G_2 and relevant generators a_1 and a_2. Bilinear group of Composite order are groups with an efficient bilinear map where the group order is a product of two primes. Assume that (G_1, G_2) are the linear assumption. The scheme utilizes the bilinear map $e : G_1 \times G_2 = G_T$ with the secure hash function $H : \{0,1\}_n^*$. Where p, q are prime numbers and $n := pq$, with $\theta(n) := (p-1)(q-1)$. The public keys are $a_1, l, z, b \in G_1$, where $a_2, d \in G_2$ such that $d := a_2^o$ where $o \in Z_n^*$ and is the private key. The proposed scheme contains the following algorithms.

- Algorithm for key generation $KG(X)$. This algorithm has one input X. The steps of the algorithm are as follows:
  1. Choose an integer b;
  2. Select two integers $f_1, f_2 \in Z_n^*$ are the secret master key;
  3. Choose two integers l, z where $l^{f_1} := z^{f_2} = b$;
  4. Choose an integer $o \in Z_n^*$;
  5. Calculate the integer $d := a_2^o$;
  6. Find to every passenger P_i a key (w_i, y_i) which are the Diffie-Hellman elements such that $1 \leq i \leq X$ as follows:
     - Choose $y_i \in Z_n^*$;
     - Calculate $w_i := a_1^{1(o||y_i)}$;

7. The key o is now a secret master key for a group key provider.

- Algorithm for signature generation $SG(pk, sk[i], m)$. This algorithm takes three parameters, public key $pk := (a_1, a_2, b, l, z, d)$, secret passenger key $sk[i] := (w_i, y_i)$ and a message $m \in \{0, 1\}^*$. Find the signature $\psi := (A_1, A_2, A_3, r, h_g, h_c, h_y, h_{a_1}, h_{a_2})$. The steps of the algorithm are:

Step 1

The description of this step is as follows:

1. Choose two integers $g, c \in Z_n$;
2. Calculate $id : (A_1, A_2, A_3) := (l^g, z^c, id, b^{g\|c})$;
3. Find the number $\partial_1 := y \cdot g$;
4. Find the number $\partial_2 := y \cdot c$;

Step 2

The description of this step is as follows:

1. Choose the integers $v_g, v_c, v_y, v_{\partial_1}, v_{\partial_2} \in Z_n$;
2. Find the numbers:

$$n_1 := l^{v_g};$$

$$n_2 := z^v;$$

$$n_3 := e(A, a_2)^{v_y} \cdot e(b, d)^{v_g - v_c} \cdot e(b, a_2)^{v_{\partial_1} - v_{\partial_2}};$$

$$n_4 := A_1^{v_y} \cdot l^{-v_{\partial_1}};$$

$$n_5 := A_2^{v_y} \cdot z^{-v_{\partial_2}};$$

Step 3

The description of this step is as follows:

1. Find the value $r := H(m, A_1, A_2, A_3, n_1, n_2, n_3, n_4, n_5)$;

Step 4

The description of this step is as follows:

1. Find the following:

$$h_g := v_g \| rg;$$

$$h_c := v_c \| rc;$$

$$h_y := v_y \| ry;$$

$$h_{\partial_1} := v_{\partial_1} \| r\partial_1;$$

$$h_{\partial_2} := v_{a_2} \| r\partial_2;$$

Step 5

The description of this step is as follows:

1. The result is $\psi := (A_1, A_2, A_3, r, h_g, h_c, h_y, h_{\partial_1}, h_{\partial_2})$;

- Algorithm for signature verification $SV(pk, m, \psi)$. This algorithm takes three parameters public key $pk := (a_1, a_2, b, i, z, d)$, message m with a signature $\psi :=$ $(A_1, A_2, A_3, r, h_g, h_c, h_y, h_{\partial_1}, h_{\partial_2})$ check that ψ is the authorized signature.
  1. Re-calculate as follows:

$$n_1' := l^{v_g}/A_1^r;$$

$$n_2' := z^{h_c}/A_2^r;$$

$$n_3' := e(A_3, a_2)^{h_v} \cdot e(b, d)^{-h_g - h_c} \cdot e(b, a_2)^{-h_{\partial_1} - h_{\partial_2}} \cdot (e(A_3, d)/e(a_1, a_2))^r;$$

$$n_4' := A_1^{h_v}/l^{h_{\partial_1}};$$

$$n_5' := A_2^{h_v}/z^{h_{\partial_2}};$$

  2. Verifies that:

$$r \equiv H(m, A_1, A_2, A_3, n_1', n_2', n_3', n_4', n_5');$$

- Algorithm for tracing signature $TS(pk, msk, m, \psi)$. This algorithm takes four parameters and its duty to trace the signature within the group. The group director is an owner of msk the master key. Also, he knows the strong Diffie-Hellman elements (w_i, y_i). Then, provided public key $pk := (a_1, a_2, b, l, z, d)$, master secret key $msk := (f_1, f_2)$, message m with a signature $\psi := (A_1, A_2, A_3, r, h_g, h_c, h_y, h_{\partial_1}, h_{\partial_2})$. Then, he can retrieve a passenger id using $id := A_3/(A_1^{f_1}, A_2^{f_2})$. Also, when a director knows the parameter w_i of passenger secret keys, he can search in the passenger list using id retrieve from the digital signature.
- Algorithm for linkable signatures $LS(pk, sk[i], m)$. This algorithm is used in certain cases for linkable signatures of the same passenger. This algorithm has three parameters the public key pk, the secret passenger key $sk[i]$ and the message m. Find the signature ψ. The steps of the algorithm are as follows:
  1. Apply the $SG(pk, sk[i], m)$:
     Encrypt the passenger id as follows:

$$(A_1, A_2, A_3) := (l^g, z^c, id.b^{g||c});$$

     Sign the message m as follows:
$$\psi := (A_1, A_2, A_3, r, h_g, h_c, h_y, h_{\partial_1}, h_{\partial_2}), \text{ with}$$
$$r := H(m, A_1, A_2, A_3, n_1, n_2, n_3, n_4, n_5) \in Z_n;$$
  2. Apply the $LS(pk, sk[i], m)$:
     Encrypt passenger id as in the above

$$(A_1, A_2, A_3) := (l^g, z^c, id.b^{g||c});$$

     Sign the message m' as follow:

$$\psi' := (A_1, A_2, A_3, r', h'_g, h'_c, h'_y, h'_{\partial_1}, h'_{\partial_2}) \text{ such that}$$

$$r' := H(m', A_1, A_2, A_3, n'_1, n'_2, n_3, n'_4, n'_5) \in Z_n;$$

Note: it can be provable that some signatures are yielded via the same passenger, since (A_1, A_2, A_3) is public. Furthermore, arbitrary integers $(v_g, v_c, v_y, v_{\partial_1}, v_{\partial_2}$ should be dissimilar compared with former one, namely $(v'_g \neq v_g, v'_c \neq v_c, v'_y \neq v_y, v'_{\partial_1} \neq v_{\partial_1}, v'_{\partial_2} \neq v_{\partial_2})$, in an attempt to undisclosed information.

- Algorithm for verification linkable $VL(\psi, \psi')$. This algorithm needs two signatures as input

$$\psi := (A_1, A_2, A_3, r, h_g, h_c, h_y, h_{\partial_1}, h_{\partial_2});$$

$$\psi' := (A'_1, A'_2, A'_3, r', h'_g, h'_c, h'_y, h'_{\partial_1}, h'_{\partial_2});$$

The result either yes or no relying on if a signature was produced via hiding the identity of the site itself (A_1, A_2, A_3) as $A_1 \equiv A'_1, A_2 \equiv A'_2, A_3 \equiv A'_3$).

5 The Proposed Scheme

The proposed scheme based on distance determines the fee that should be paid under an algorithm designed for this purpose from the point of entry to the exit point. Also, a scheme must contain the entry station *id* in the entry ticket, and it must determine the validity duration of every ticket. In addition, create relationship between the distance that the passenger wants to reach and the fees that must be paid, in case the distance traveled is increased, the ticket price is increased.

The entry ticket should contain another important element that is the destination of the passenger's trip. It is verified by a direction mentioned in the entry ticket to benefit from avoiding engaging in conversation attacks when the scheme wants to meet its two demands. First, in a departure station passengers should receive a multi-point entry ticket in accordance with their destination station so that there are separate entrances in the direction. Second, the passenger is unable to alter the movement of direction without leaving a scheme. Once the passenger reaches a stop station, they should leave by its direction. It means that the exit scheme is separated via direction.

When a service is not compatible with the demands mentioned above, there is a need for a more complicated scheme to resolve issues that may arise because of the probability of conversation attacks. Full description to resolve solution of compatible services is illustrated in Sect. 6.1 whereas solution of incompliant services is explained in Sect. 6.4.

5.1 The Properties of the Proposed Scheme

The following characteristics are necessary to design the proposed scheme:

- Anonymity: the scheme should secure pseudonym to meet the passenger acceptance. The proposed scheme supports non-pseudonym for monitoring and security purposes. So, the compromise solution is the revocable anonymity. It means that when the passenger cheats, his pseudonym is canceled.
- Authenticity: each ticket should be issued through the authorized service provider.
- Integrity: once a ticket has been issued, it may not be amended.
- Un-repudiation: a provider cannot repudiate an issue of any ticket.
- Validity time: each issued ticket holds the validity time indicator, to verify if it is valid or not. Also, every ticket must remain issued in the database till its validity period is ended.
- Un-traceability: the authorized service provider can trace the passenger's entry and exit, but does not allow him to trace various trips for the same passenger.
- Double-spending: The validity time of each ticket must be verified. If the checking is true, the scheme verifies a ticket has not on a database for consumed tickets via its serial number. Such checking guarantees that a ticket has not been utilized more than once.

5.2 Scheme Participants

The following participants play the main role in the scheme proposed:

- Passenger P: enters the scheme and pays the ticket fare at checkout. This can be done by using their mobile phone.
- Authorized service provider: The responsibility of his work shall be at the entry and exit stations and his duty is to check the ticket in the possession of the passenger and calculate the ticket price that should be paid by the passenger in accordance with the element determined in the algorithm.
- Trusted authority TA_1: administer the amounts paid by travelers once they are leaving a scheme.
- Trusted authority TA_2: administers a passenger revocation list and the set keys and also he can cancel the pseudonyms of the passenger when he cheats.

5.3 Scheme Description

In the proposed scheme, there are four protocols which are as follows:

1. Initialization Protocol

This protocol is implemented only once at the start and ends its mission. The trusted authority TA_2 implements KG algorithm with (X) input, to yield a collection of length $pk, sk[i], rl[i], g, p, q)$, such that pk is a public group keys, $sk[i]$ is a secret key to every passenger P_i, $rl[i]$ is a revocation list, and (g, p, q) are public keys, such that g is an encryption key and p, q are primes such that $n = pq$. Also, every authorized service provider ticket selects his public and private keys. The secret group keys $sk[i]$ are published if the passenger is registered in the trusted authority TA_2.

2. Passenger Registration Protocol

The passenger P should register with the trusted authority TA_2 to receive his public-key and private-key $(pk, sk[i])$. The passenger should consent that his id will be revealed if he defrauds, or when the court wants to revoke his pseudonym. In addition, the passenger P should register with the trusted authority TA_1 by pseudonym; this is used just for the payments. The trusted authority TA_1 is then set up by the passenger and authorized service provider, to deal with the payment, and to ensure that a payment is a legitimate transaction according to scheme requirements. The passenger creates anonymous name id_p for payment and selects an arbitrary integer $j_p \in Z_q$. Such anonymity name id_p must be proved by the trusted authority TA_1 using Feige et al., zero-knowledge proof of identity [7] without revealing the information of j_p. Therefore, the privacy of the passenger is protected. However, such information can be revoked via the trusted authority TA_2 when needed. The registration protocol is described as follows.

Algorithm of Anonymity Generation: Passenger P should do the following
1. Select an arbitrary integer $j_p \in Z_n$ // represent passenger payment private
2. Find the value $id_p = g^{j_p} \bmod m$;
3. Pass his id_p and the signed message $\Omega_p(id_p, \text{'}hi\text{'})$ to a trusted authority TA_2;

Algorithm of Key Issuing: Trusted authority TA_2 should do the following:
1. pass the key pair $(pk, sk[i])$ to the passenger P;
2. pass the public keys (a, p, q) to the passenger P;
3. pass the signed message $\Omega_{TA_2}(id_p)$ to the passenger P;

Algorithm of Beginning Zero Knowledge Protocol: Passenger P should do the following:
1. select an arbitrary integer $g_o \in Z_n$;
2. find $g_1 := g^{g_o} \bmod n$;
3. pass $(id_p, g_1, \Omega_{TA_2}(id_p))$ to the trusted authority TA_1;

Algorithm of Challenge Generation: Trusted authority TA_1 should do the following:
1. select the challenge integer $g_2 \in Z_n$;
2. pass g_2 to the passenger P;

Algorithm of Proof Generation: Passenger P should do the following:
1. Calculate the zero knowledge protocol $g_3 = g_o || g_2 \cdot j_p \bmod n$;
2. pass g_3 to the trusted authority TA_1;

Algorithm of Check Anonymity: Trusted authority TA_1 should do the following:
1. check the $g^{g_3} \equiv g_1 (id_p)^{g_2}$t $g^{g_3} \equiv g_1 (idp)^{g_2}$;

3. Scheme Entry Protocol

If the passenger P reaches to the entry station es, he must enter the scheme rightly. Then, he will obtain the entry ticket e_{in} and according to this ticket he can pay the ticket fare. However, when he enters into the scheme, he should generate the group signature that confirms he is one from the authorized group of passengers, whilst his id is not revealed. Then, such signature is passed to the authorized service provider in the entry station es, and then obtains the entry ticket e_{in} from entry station es. Finally, when the passenger arrives at his destination, he should show this ticket to the destination station ds. The scheme entry protocol is described as follows:

Algorithm for Obtain Service: Passenger P should do the following:
1. Select a random integer $y_1 \in Z_n$;
2. Find $y_2 := g^{y_1} \bmod n$;
3. Find $w := pk_{TA_1}(id_p)$; // Probabilistic encryption for id_p passenger payment pseudonym
4. Select a random integer $u \in Z_n$; // verification element
5. Find a secure hash function of u as follows $x_u := H(u)$; // x_u hash image for parameter u
6. Form $\psi := (y_2, w, x_u)$;
7. Sign ψ as follows $\ddot{\psi} := (\psi, \bar{\psi} = SG(pk, sk[i], \psi))$; // Using algorithm for signature generation.
8. Pass $\ddot{\psi}$ to the authorized service provider in the entry station es;

Algorithm for Generating Ticket: The authorized service provider in es should do the following:
1. Checks a signature $\ddot{\psi}$, if a signer is the valid passenger group member by using the algorithm for signature verification as follows $VG(pk, \psi, \bar{\psi})$;
2. Select an entry timestamp, it means the time login scheme t_1;
3. Find the entry ticket $en := (s, es, t_1, t, \ddot{\psi}, \lambda)$; // where s is a serial number of the ticket issued by authorized service provider in es, where λ is a trip direction and t a time validity.
4. Sign an entry ticket $e'_{in} := (e_{in}, \Omega_{es}(e_{in}))$;
5. Pass the signature of entry ticket e'_{in} to the passenger P;

Algorithm for Check Entry: passenger P should do the following
1. check a signature of entry ticket e'_{in} and its information;

4. Scheme Exit Protocol

If the passenger arrives at his destination station ds he should show his entry ticket e_{in} to checkpoint. Then, the authorized service provider at checkpoint computes the charges due. The passenger must consent for the charges calculated, and then passes this information in addition to the receipt as evidence that sends them by a pseudonym and privately to the trusted authority TA_1.

Only the trusted authority TA_1 knows such information, also the destination station ds is not allowed to reveal the content. Finally, the trusted authority TA_1 records the cost into the passenger P account. When the entire operation goes correctly, the passenger P obtains a ticket to exit; this is the confirmation which shows that passenger P followed the procedures accurately. The scheme exit protocol is described as follows:

Algorithm for Proof Ticket: Passenger P should do the following:
1. Encrypt the verification key u;
2. Pass $(e'_{in}, pk_{ds}(u))$ to the authorized service provider in an destination station ds;

Algorithm for Checking Ticket: The authorized service provider in ds does the following:
1. Check signature of entry ticket e'_{in} that is calculated by the authorized service provider in es;
2. Check that $\psi \cdot x_u \equiv H(u)$; // This confirms that a passenger P is the true owner of the ticket e_{in};
3. Check that $e_{in} \cdot s$ // s is serial number posted via passenger and is not used before;
4. Select a validity time t is not expired, and a trip direction λ;
5. Select a timestamp t_2 where $t_2 \geq t_1$;
6. Find the charges $f := (e_{in}.es, d, e_{in}.t_1, t_2)$; // where f is required fee
7. Select a challenge $y_4 \in Z_n$;
8. Compute $c := (e'_{in}.u, f, y_4, t_2, ds)$;
9. Sign the challenge $c' := (c, \Omega_{ds}(c))$;
10. Pass the signature c' to the passenger P;
11. Calculate $o_{ds} := (c \cdot f, e_{in}.s, e_{in}.\psi, y_4)$;

Algorithm to Establish Payment: The passenger P should do the following:
1. Check the signature of c' that is calculated by destination station ds;
2. Find the Feige et al. zero knowledge protocol $y_5 := y_1 || y_4 \cdot j_p \bmod n$;
3. Encrypt the pseudonym of payment $id_p := pk_{TA_1}(y_5, e_{in} \cdot s, s \cdot f)$;
4. Pass the pseudonym of payment id_p to a destination station ds;

Algorithm for Passing Payment Information: The service provider in ds does the following:
1. Repost the pseudonym of payment id_p to the trusted authority TA_1;
2. Resend the verification information o_{ds} to the trusted authority TA_1;

Algorithm for Checking Payment: Trusted authority TA_1 should do the following:
1. Retrieve anonymity of payment id_p to find the Feige et al., proofs of identity y_5;
2. Recover $e_{in}.\psi \cdot s$ to find the pseudonym id_p;
3. Withdraw the money from the passenger account;
4. Check id of passenger P using Feige et al. zero knowledge protocol by $f^{y_5} \equiv y_2 \cdot (id_p)^{y_4}$;
5. If true, the fee f is withdrawn from the passenger account id_p.
6. If not, compute the payment refusal $pr := (id_p, 'refuse')$;
7. Sign the payment refusal as follows $pr' := (pr, \Omega_{TA_1}(pr))$;
8. Pass a payment refusal signature pr' to destination station ds and ends the algorithm

9. Compute the payment acceptance $pa := (e_{in}.s, c \cdot f, 'accept')$;
10. Compute the signature of pa as follows $pa' := (pa, \Omega_{TA_1}(pa))$;
11. Pass payment acceptance signature pa' to authorized provider in the destination station ds;

Algorithm for Collection Exit: The provider in the destination station ds should do the following:
1. Compute the exit ticket $e_{out} := (e_{in}.s, ds, c \cdot f, c.t_2)$; // leave at $b.t_2$
2. Compute the signature of exit ticket e'_{out} as follows $e'_{out} := (e_{out}, \Omega_{as}(e_{out}))$;
3. Pass the signature of the exit ticket e'_{out} to the passenger P and let him exit from the scheme;

Algorithm to Verify Ticket: The passenger P should do the following:
1. Check the signature of the exit ticket e'_{out};

5.4   Passenger Assumptions

In the scheme exit protocol, the issuer in the destination station ds cannot trace the protocol because of various causes. For example, the issuer in the destination station ds might make errors, carry out scheme crash, make an immoral act, or execute criminal behavior. Due to these reasons, the true passenger will get inappropriate care. To resolve such troubles, the proposed exit protocol may confront two passenger assumptions.

1. Received a passenger signature on fare c' invalid
In the scheme exit protocol, the passenger P passed valid information, which is the signed entry ticket and the verification key (e'_{in}, u) to the destination station ds, but the issuer in the destination station ds might mistreat and pass incorrect information concerning fare c' such as, sent inexact time exit ticket t_2 to passenger P. In this case, the passenger P can use the information received from the destination station ds as evidence when he communicates with the trusted authority TA_1 to demonstrate that he has the right ticket to leave via using the following steps.

Step 1 request: The passenger P does the following:
1. Repost the signed entry ticket (e'_{in}, u) to the trusted authority TA_1;
2. Repost the incorrect information c' to the trusted authority TA_1;

Step 1 reply: The payment trusted authority TA_1 does the following:
1. Check the signature of entry ticket e'_{in} is calculated by authorized service provider in the es;
2. Check that the hash image $\psi \cdot x_u \equiv H(u)$. This confirms that the passenger P is the true owner of the ticket e_{in};
3. If the passenger fare c' is wrong then
   The trusted authority TA_1 should check that $c.t_2$ or $c.f$ if are not correct. For instance, the present time is $< c.t_2$;
4. Check that the time validity t is valid
5. Check that the direction trip λ is true;

6. Select a new timestamp t_2
7. Compute the charges that should paid according to the differences between the entry time and the exit time (t_1, t_2) by $f := (ds, e_{in}.es, e_{in}.t_1, t_2)$;
8. Select an integer $y_4 \in Z_n$;
9. Compute $c := (e'_{in}, f, y_4, t_2, ds)$;
10. Sign $c' := (c, \Omega_{ds}(c))$;
11. Pass c' to the passenger P;

Step 1 Remark: The other steps of the exit protocol will continue normally

2. Received a wrong signature on exit ticket e'_{out}

In the scheme exit protocol, the passenger P passed valid information, which is the signed entry ticket, the verification key and his identity (e'_{in}, u, id_p) to the destination station ds, but the authorized service provider in the destination station ds might mistreat and pass incorrect information concerning the exit ticket e'_{out} fare c' to passenger P. In this case, the passenger P can use the information received from the destination station ds as evidence when he communicates with the trusted authority TA_1 to demonstrate that he has the right ticket to leave via using the following steps.

Step 2 request: The passenger P does the following:
1. Repost (e'_{in}, u, c', id_p) to the trusted authority TA_1;

Step 2 reply: The trusted authority TA_1 does the following:
1. Check the signature of entry tick e'_{in} that has been calculated via a destination station ds;
2. Check id of the passenger P using Feige et al., zero knowledge protocol of identity as follows $f^{y_5} \equiv y_2(id_p)^{y^4}$;
3. Check that the hash image $\psi \cdot x_u \equiv H(u)$. This confirms that the passenger P is the true owner of the ticket e_{in};
4. Verify that a time validity t is still valid;
5. Verify that the trip direction λ is true;
6. Compute the charges that should paid according to the differences between the entry time and the exit time (t_1, t_2) by $f := (en.es, c, ds, en.t_1, c.t_2)$;
7. Check the fare f is equivalent to $c.f$ if not, the passenger P should fix it according to the protocol in the assumption 1;
8. Compute the exit ticket $e_{out} := (e_{in}.s, c.f)$;
9. Sign the exit ticket $e'_{out} = (e_{out}, \Omega_{TA_1}(e_{out}))$;
10. Pass the signature of the exit ticket e'_{out} to the passenger P;

Step 2 Remark: The other steps of the exit protocol will continue normally.

Notes

In both assumptions, the trusted authority TA_1 should:
1. Inform the provider in the destination station ds about his bad behavior with passenger.
2. Inform the provider in the destination station ds and warn him that more actions will be taken against him when such problems continue.

5.5   Authorized Service Provider Assumptions

In the scheme exit protocol, the passenger P cannot trace the protocol because of various causes. For example, the passenger P might make errors, carry out scheme crash, make an immoral act, or execute criminal behavior. Due to these reasons, the authorized service provider will get inappropriate care. To resolve such troubles, the proposed exit protocol may confront two issuer's assumptions.

1. Received invalid signature of entry ticket and verification key (e'_{in}, u)

In a scheme exit protocol, the authorized service in the destination station ds gets the signature of entry ticket and the verification key (e'_{in}, u) for checking and finds that information is not true. In this case, he could reveal the passenger id_p via using the following steps:

Step 3 request: The authorized service provider in the destination station ds does the following:

1. Pass (e'_{in}, u) to the trusted authority TA_2;

Step 3 claiming passenger: The passenger p does the following:

1. Pass the same information (e'_{in}, u) to the trusted authority TA_2, to prevent untrue claims;

Step 3 reply: The trusted authority TA_2 does the following:

1. Check the signature of (e'_{in}, u) that created via departure station es;
2. Check the concurrent with the hash image $e_{in}.\psi \cdot x_u \equiv H(u)$;
3. If the concurrent is not true, the trusted authority TA_2, frustrate the request;
4. Check that the set signature of $e_{in}.\ddot{\psi}$ is created via the passenger P. Then, detect who is the signer within the group;
5. Pass the passenger id_P to the authorized service provider in the destination station ds;
6. Send the passenger pseudonym for payment id_p to the trusted authority TA_1;
7. Enter the passenger id_p into a revocation list;

2. Received a wrong pseudonym passenger for payment id_p

In the scheme exit protocol, the provider in destination station ds and the trusted authority TA_1 gets the checking step of the passenger pseudonym id_p. But, such information may not be true. Thus, the provider in the destination station ds could claim to detect passenger id_p via using the following steps:

Step 4 request: The trusted authority TA_1 does the following:

1. Compute the fare rejection $pr := (id_p)$; // means checking information fault
2. Sign the fare rejection $pr' := (pr, \Omega_{TA_1}(pr))$;
3. Pass the signature of the fare rejection pr' to the destination station ds;
4. Pass $(sk_{ds}(id), o_{ds})$ to the trusted authority TA_2;
5. Terminate the protocol;

Step 4 authorized service provider information: The provider in the ds does the following:

1. Pass (e'_{in}, u) to the trusted authority TA_2;

Step 4 claiming passenger: The passenger P does the following:

1. Pass (e'_{out}, u, id_p) to the trusted authority TA_2; //to prevent untrue claims

Step 4 reply: The trusted authority TA_2 does the following:

1. Check that a recovered message of id_p, o_{ds} with (e'_{in}, u) are connected;
2. Check that the set signature of $en_{in}.\ddot{\psi}$ is created via the passenger P; // to detect who is the signer within the group;
3. Pass the passenger id_p to the authorized service provider in the destination station ds;
4. Pass the passenger pseudonym payment id_p to the trusted authority TA_1;
5. Enter the passenger id_p to a revocation list;

6 Scheme Analysis

The proposed online scheme achieves the security properties described in Sect. 6.1. The achievements are as follows:

6.1 Authenticity

The thought of forged tickets is really hard to make these days. From another side, entry ticket is signed by $e'_{in} := (\Omega_{es}(e_{in}))$; exit ticket is signed by $e'_{out} := (e_{out}, \Omega_{ds}(e_{out}))$, and posted information previous to a payment $c' := (c, \Omega_{ds}(c))$. But, when the dishonest passenger can generate a legitimate ticket without known the secret keys to neither departure station es nor the destination station ds. Signatures can be created during the impersonation of such issuers. But, if we employ safe digital signature scheme, such process is considered to be useless. From another side, the passenger P passes the checking message signed by his group secret key $\ddot{\psi} := (\psi, Sign_G(\psi))$. Also, such signature ensures that the information is real and is published by the true passenger P.

6.2 Non-repudiation

It is impossible for the issuer to repudiate the issued ticket, signed through an authorized authority. Assuming that the signing of a secure passenger scheme, the process can be executed only via such issuers. Therefore, an issuer id is connected with the signature algorithm of the ticket. So, it is impossible for the issuer to deny its possession. The same thing happens with the group signing scheme, if id is revealed, the owner of the message can be checked.

6.3   Integrity

Once the ticket is issued, there is no way for it to modify. However, if a secure hash function is used, its inverse is hard to calculate. But, when the ticket information is altered, the signature validation process will be improper. To pass a checking, a signature requires to be recreated again for the new ticket information. But, such process is practically useless at present with most present devices. The same thing happens for the group signing scheme.

6.4   Validity-Time

The ticket will not be accepted when its validity time is ended. The issuer in the destination station ds takes the ticket from a passenger for inspection. In such checking, the issuer should detect the differences between the present time and the validity time of signature entry ticket e'_{in} that is signed via departure station es.

6.5   Anonymity

In this part, two types of anonymities are discussed, these are as follows:

1. The Ticket Anonymity: The id_p content is encrypted by a trusted authority TA_1 exponent key. The issuers (in the departure station es and in the destination station ds cannot access to such content since it requires the secret key from the trusted authority TA_1. In the proposed online scheme, the passenger P calculates the group signature $(e_{in}.\ddot{\psi} := (\psi, Sign(\psi)))$, this confirms that a signer is an authorized member in the group signature. When testing the characteristics of the group signing scheme, the issuer cannot determine the signing generator. If there is an argumentative case, it is possible to detect the passenger who signed the message by the collaboration between the trusted authority TA_1 and the trusted authority TA_2. In case the passenger shows in a revocation list, he is identified, allowing more action against him.

2. The Passenger Anonymity: The content regarding the fare is encrypted by the trusted authority TA_1 exponent key, and just the trusted authority TA_1 can entree to that information. The issuer is out of payment and he is receiving the fee confirmation of trusted authority TA_1 only. Therefore, the trusted authority TA_1 has id_p information regarding (j_p, id_p) such that $id_p := f^{j_p} \bmod p$ that recognizes him as the true passenger. Thus, validate the passenger via verifying the content of j_p by Feige *et al.* zero knowledge protocol.

6.6   Traceability

The group signature scheme used in the online proposed scheme [7] by Bellare-Neven employs a random oracle model under the RSA. In this case, it is impossible to predict encrypted message because of a particular message. This makes it impossible to trace between various group signatures made via a same passenger. Note that group signature made via a same passenger should be undetectable by the authorized issuers in the departure station es and in the destination station ds).

6.7   Double-Spending

In case the passenger attempts to use the entry ticket as a double-spending, the answer will be that the serial number of that ticket is already used. When this passenger misconduct is proven, the trusted authority TA_1 can add such passenger to the revocation list. Thus, the proposed online scheme prevents ticket double-spending.

6.8   Fraud Attack

The scheme is impossible to attack via the fraud passengers. Because, such an attack depends on exchange of entry tickets and does not apply on time-spend schemes, because passengers do not get any utility from such exchange. Charges are computed employing the time-stamp of entry ticket and when passengers exchange the tickets, fees remain the same, and one of the travelers will bear more than his actual ticket. Therefore, passengers are frustrated from exchange tickets.

6.9   Remarks

It is impossible to attack the scheme based on distance explained in Sect. 6 through passengers who engage in conversations among themselves in order to penetrate the scheme by exchanging the tickets with each other or manipulating the ticket prices. In the scheme based on distance, any attack according to ticket exchange will only be useful in certain situations. In each of such situations the direction of passengers should be different. When the scheme depend on distance effectively meets separate needs and directions, passengers traveling in opposite directions will not be capable to use an exchanged ticket to exit the scheme because the direction in the ticket will be different from the actual ticket of the passenger.

7   The Application

The application of the system has been done as to gauge its performance. The purpose of this system is for utilizing it through Smartphone systems. So, it must be executed on a portable scheme. Nowadays, there are some portable systems on hand. Nevertheless, the Android operating system appears to be increasing within the entire range of Smartphones [8], from medium-level to advanced-level. In addition, this operating system utilizes the C++ and Java programming languages for implementations evolution and maintained via Google database.

A headmost challenge is the application of "an identity-based multi-signature from RSA" introduced by Bellare et al., [7]. After conducting the profound search into a web, it was not possible to get any application from the C++ pairing-typed system. Thus, chose to evolve the application from zero point. Since the system uses the pairing-based, it was decided for utilizing the C++ bilinear mapping encryption library which remains the entire C++ executor from the bilinear-typed encryption library [9]. The C++ pairing-based encryption library gives a capability to calculate the complicated conjugations across the techniques needed, through multi-signature systems, as the multi-signature utilizes the encryption by bilinear mapping. Therefore, and for uniting the operations, it was chosen the calculations processes as well as modular exponentiations, which such type of an encryption are required. Moreover, a digital signature as well as encrypted data algorithms utilizes the RSA system, which uses in turn the bouncy castle library [10]. This library is a set of APIs utilized in cryptography. It also includes APIs to C# language. However, the applications are divided into two main divisions, namely a user part and a server part. This can be essential to make connections between every side that is designed into the Extensible Markup Language (XML). However, the summarized description of each part is as follows:

- User part: by means mentioned above, a passenger implementation P was designed by Android OS based on 4.4 API Android Developers for matching with the mobile phones.
- Server part: such part considers the servers that related to the trusted authority TA_1, trusted authority TA_2, an entry station, as well as the server related to the destination station. Every one of these is programmed by C++ Development Kit language Windows version 10. Also, each of which has its database for retaining beneficial data and displays its services by the *Transmission Control Protocol* (TCP) executor from a server.

7.1 Test Phase

In this phase, the laptop has been utilized. In the above, the information needed and related to the servers is mentioned. A user part was tested in three types of mobile devices. Firstly, Samsung Galaxy A7, secondly Huawei Mate 10 and thirdly HTC Desire 10 Pro. These devices are from the medium-class Smartphones. The properties of these Smartphones were listed on Table 1.

The implementation test on the three kinds of mobile devices could supply useful knowledge concerning the system performance across three kinds of mobiles that have dissimilar running abilities. This is furthermore important to consider about the next generations of Smartphones, and the predictable enhancement in conduct by utilizing modern and robust devices. A test was performed for every mobile phone with every algorithm ten times for obtaining a mean time to every step from the steps of the algorithm. Lastly, it is important to implement connection amidst the operating system Android uses with the proposed system that is offered via the 802.11 g wireless net technology. Nevertheless, the communications among every infrastructure server remain performed on a computer LAN interface.

7.2 Discussion

Following execution of the proposed system come the discussions then the analysis of outcomes to every test. Then, it would notice the metrics outcomes with all specifics. In Fig. 1 it sees that the carrying out of the system on Samsung Galaxy A7 remains quicker compared by Huawei Mate 10 and by HTC Desire 10 Pro as predictable. On the other hand, when every algorithm in the proposed system is parsed separately the Samsung Galaxy A7 requires around 22.2 s for implementing the proposed system in

Table 1. The smartphones test properties.

Feature	Device		
	Samsung Galaxy A7 Mobile phone	Huawei Mate 10 Mobile phone	HTC Desire 10 Pro Mobile phone
CPU	1.9 GHz Octa-core	2366 MHz Octa-core	2000 MHz Octa-core
RAM	3 GB	4 GB	4 GB
ROM	32 GB + Micro SD	64 GB-Micro SD	64 GB + Micro SD
Biometrics	Fingerprint (touch)	Fingerprint (touch)	Fingerprint (touch)
OS	Android 8.0	Android 8.0	Android 6.0
Dimensions	(156.8 × 77.6 × 7.9 mm)	(150.5 × 77.8 × 3.2 mm)	(156.5 × 76.0 × 7.86 mm)
Weight	(168 g)	(186 g)	(165 g)
IP certified	IP 68	IP 68	IP 68
Resolutions	1080 × 1920 Pixels	1440 × 2560 Pixels	1080 × 1920 Pixels
Flash	LED	Dual LED	Dual LED
Camera	16 megapixels	12 megapixels	20 megapixels
Storage expansion	Up to 256 GB	Up to 256 GB	Up to 2000 GB
Display	Full HD	16:9	5.5-inch diagonal
Battery	3600 mAh	4000 mAh	3000 mAh
Release date	Jan 13, 2017	Oct 20,2017	Oct 1,2016

full, comparing to Huawei Mate 10 which requires about 32.5 s, whilst HTC Desire 10 Pro needs around 40.8 s and this means it takes about 4.6 times more comparing to the Samsung Galaxy A7. However, it could be mentioned that the dissimilarity among the three Smartphones is not great. But, for display a report about taken time and to the way of the enhancement an implementation performance of all mobile devices, it is important to get a time spent to implement every step of the algorithm steps and to consider the implementations which need more processes calculation and operations exponentiation.

The implementation test on the three kinds of mobile devices could supply useful knowledge concerning the system performance across three kinds of mobiles that have dissimilar running abilities. This is furthermore important to consider about the next generations of Smartphones, and the predictable enhancement in conduct by utilizing modern and robust devices. A test was performed for every mobile phone with every algorithm ten times for obtaining a mean time to every step from the steps of the algorithm. Lastly, it is important to implement connection amidst the operating system Android uses with the proposed system that is offered via the 802.11 g wireless net technology. Nevertheless, the communications among every infrastructure server remain performed on a computer LAN interface.

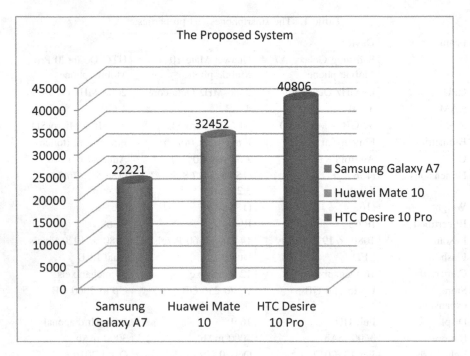

Fig. 1. Overall time in milliseconds disbursed to the proposed system and on every mobile device separately.

The Fig. 2 illustrated a time inflow of proposed system and mobile devices. For knowing the Fig. 2, it requires to describe the meaning of every step of x-axis:

- Download bilinear-map: A passenger implementation loads entire information that is necessary for performing the pairing processes. For example, insert elliptic curve keys from the file and setting up chosen C++ objects.
- RSA Setup: RSA public and private keys are inserted from the Privacy Enhanced Mail (PEM) document kept at Smart mobile storage. For example, of Secure Digital (SD) or build-in memory.
- Public key uses: Time required in implement the requirement of trusted authority TA_2 server of an encryption key f that is required to be used on a next step.
- Registration of the trusted authority TA_2: The implementation should be accomplishing by the registration with the trusted authority TA_2 server. So that it corresponds with the steps of Algorithm for anonymity generation and with steps of the Algorithm for key issuing as in 6.3.2.
- Registration of trusted authority TA_1: It includes a passenger registration that is a part of the trusted authority TA_1 server. So that it corresponds to algorithm steps that begin with zero knowledge protocol and with steps of the Algorithm for challenge generation, as well as with steps of the Algorithm for proof generation also with steps of the Algorithm for verify pseudonym as in 6.3.2.

- Entrance set up: It includes an overall pre-calculation time expended prior to entering the scheme. Now, all pre-calculable keys required to construct a multi-signature is founded.
- Entering: It includes a time it takes for entering the system. This corresponds to the algorithm steps of 6.3.3.
- Exit set up: It includes a time it takes via exit multi-signature before the calculation and prior to an exit step.
- The exiting: It includes a spent time from system exiting. This corresponds to the algorithm steps of 6.3.4.

However, the Fig. 2 displays the time inflows of proposed system. But, when the chart is parsed, it could notice that the majority of time is spent for calculating entrance algorithm, it takes about (9733, 13777, 16606) milliseconds when executing in Samsung Galaxy A7 and Huawei Mate 10 and HTC Desire 10 pro) respectively. So, it will be obvious that the use of Smart phone Samsung Galaxy A7 type is quicker especially when it is applied to the entrance algorithm merely. So, it prefers to use the pre-calculation prior to use in the entrance algorithm, in order to remain a system performance comparable across all Smartphones.

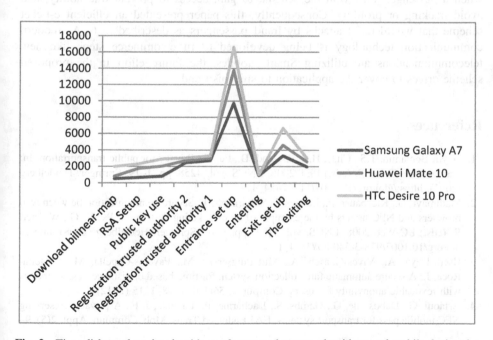

Fig. 2. Time disbursed on the algorithms of proposed system algorithms and mobile devices by milliseconds.

However, an orientation remains similar until the entrance algorithm, since the preceding steps remain not affected. Then, prior to leaving a system, another leaving step shows. This requires that a passenger should make additional advance calculations before reaching an exit set-up algorithm. However, the exit set-up algorithm has implemented through Samsung Galaxy A7 firstly and took about 3.3 s whilst in Huawei Mate 10 took around 4.6 s and implemented lastly on HTC Desire 10 Pro and took about 6.7 s. It means that, it needs for 3.4 s extra for less modern mobiles. However, when it compares by every mobile device separately, it could notice that the time required for implementing the adjusted version, by not considering for the pre-calculation period that could calculated offline with the passenger not more of a main time consumed. Thus, it could say that the adjusted version raises a cost of time nonetheless and stays valid to all mobiles.

8 Conclusions

This paper introduced a secure e-ticketing scheme that is convenient to large-scale public transfer. It accomplished a scheme which is appropriate for distance-type prices with minor alterations. Using group signature schemes, it is possible to authenticate the passenger whilst maintaining privacy. But, passenger *id* can be detected if the passenger misconducts. The proposed scheme does not need a new permission each time when a passenger joins with the scheme to gain access to prevent traceability, and avoid tracking or profiling. Consequently, this paper presented an efficient e-ticket scheme that would resist attacks by fraud passengers as described in 6. As modern communication technology is being developed to fit e-commerce steps like new telecommunications are utilizing Smart mobiles, the future effort in the proposed scheme drives to cover the application to suit this trend.

References

1. Heydt-Benjamin, T.S., Chae, H.-J., Defend, B., Fu, K.: Privacy for public transportation. In: Danezis, G., Golle, P. (eds.) PET 2006. LNCS, vol. 4258, pp. 1–19. Springer, Heidelberg (2006). https://doi.org/10.1007/11957454_1
2. Madlmayr, G., Kleebauer, P., Langer, J., Scharinger, J.: Secure communication between web browsers and NFC targets by the example of an e-Ticketing system. In: Psaila, G., Wagner, R. (eds.) EC-Web 2008. LNCS, vol. 5183, pp. 1–10. Springer, Heidelberg (2008). https://doi.org/10.1007/978-3-540-85717-4_1
3. Ilsern-Deyà, A., Vives-Guasch, A., Mut-Puigserver, M., Payeras-Capella, M., Castella-Roca, J.: A secure automatic fare collection system for time-based or distance-based services with revocable anonymity for users. Comput. J. **56**(10), 1198–1215 (2012)
4. Arfaoui, G., Dabosville, G., Gambs, S., Lacharme, P., Lalande, J.-F.: A privacy-preserving NFC mobile pass for transport systems. EAI Endorsed Trans. Mob. Commun. Appl. **2**(5), 9–12 (2014)
5. Aboud, S., Alfyoumi, M.: Efficient e-Tickets fare scheme for time-typed services. J. Internet Bank. Commer. (1), 20 (2017)
6. Bellare, M., Neven, G.: Identity-based multi-signatures from RSA. In: Abe, M. (ed.) CT-RSA 2007. LNCS, vol. 4377, pp. 145–162. Springer, Heidelberg (2006). https://doi.org/10.1007/11967668_10

7. Feige, U., Fiat, A., Shamir, A.: Zero-knowledge proofs of identity. J. Cryptol. **1**(2), 77–94 (1988)
8. http://www.bouncycastle.org/. Accessed 2013
9. Martorell, G., Riera-Palou, F., Femenias, G.: Performance analysis of fast link adaptation-based 802.11n basic and RTS/CTS access schemes. X Jornadas de Ingeniería Telemática JITEL, pp. 182–190 (2011)
10. Source GPL v2 version of yaSSL, Version: 2.4.2. https://www.wolfssl.com/products/yassl/. Accessed 22 Sept 2016

Robust Blind Digital 3D Model Watermarking Algorithm Using Mean Curvature

Zainab N. Al-Qudsy[1(✉)] ⓘ , Shaimaa H. Shaker[2],
and Nazhat Saeed Abdulrazzque[3]

[1] Computer Sciences Department,
Baghdad College of Economic Science University, Baghdad, Iraq
zainabqudsi@gmail.com
[2] Computer Sciences Department, University of Technology, Baghdad, Iraq
120011@uotechnology.edu.iq
[3] Electro Mechanical Engineering Department,
University of Technology, Baghdad, Iraq
drnazhatsaeed2016@gmail.com

Abstract. In this study, a robust and blind 3D model watermarking algorithm is introduced based on studying the geometrical properties of the 3D model. This work shows that the mean curvature (MC) has an important feature that can be used to classify the surface points. During embedding, the vertices of the surface are classified depending on the measured MC to select suitable domain for embedding and improve the imperceptibility and robustness of the watermarking algorithm. During extraction, the watermark is extracted without the need of the original 3D model. The proposed watermarking technique solves the conflict between robustness and imperceptibility by introducing watermarking algorithm that is resistant against different types of attack without affecting the quality of the watermarked model. The performance evaluation shows high imperceptibility and tolerance against many types of attack, such as noise addition, smoothing, cropping, translation, and rotation. The proposed algorithm has been designed, implemented, and tested successfully.

Keywords: Digital 3D model watermarking · Mean curvature
Geometrical properties

1 Introduction

Three-dimensional models are widely used in computer vision and computer graphics and are deployed in various domains, such as web applications, video games, movies, and medical visualization [1]. Generally, the present Internet network provides users with high flexibility by enabling them to perform illegal manipulation operations. Watermark arises as one of the solutions to this problem. Particularly, watermarking in a 3D model is the practice of invisibly altering the model by embedding small information called "watermark," which is used as a copyright protection tool for 3D models from illegal copy, tampering, and distribution without authorization from the actual owners of these data. An efficient 3D model watermarking algorithm must provide high

© Springer Nature Switzerland AG 2018
S. O. Al-mamory et al. (Eds.): NTICT 2018, CCIS 938, pp. 110–125, 2018.
https://doi.org/10.1007/978-3-030-01653-1_7

imperceptibly and good robustness against different types of attack. In other words, a contradiction exists among achieving these requirements. To handle 3D models, multiple file formats exist for 3D models that are available for downloading over the Internet network, the most popular of which that have been used by researchers with the watermark are .obj and .off [2]. In the graphical design field, a model must be represented in a suitable structure for rendering. The typical structure is the 3D triangular mesh. In this representation, a model is defined by a set of polygons, which contain three types of element, namely, faces, edges, and vertices. Each vertex can be represented by three Cartesian coordinates X, Y, and Z, which are linked by edges to form a set of facets [1]. Triangular mesh constructed from a set of triangles is linked by their common edges with three vertices for every triangle. The 3D triangular mesh for a cow model and its zoomed-in version that explains the rings around each vertex are shown in Fig. 1.

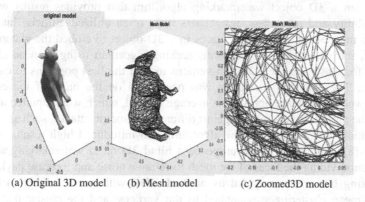

  (a) Original 3D model   (b) Mesh model   (c) Zoomed3D model

Fig. 1. 3DCow model

Numerous important have been conducted on hidden watermarks in 3D models. Garg and Suneeta presented a non-blind watermarking algorithm by dividing the vertices of a 3D mesh surface into flat, peak, and deep regions depending on the values of the mean curvature (MC). The watermark bits are embedded in deep vertices that have negative MC values based on the hash code obtained by applying MD5 hash function to the watermark. Their work is robust against various distortion attacks, such as smoothing, simplification, subdivision, and distortion-less attacks [4]. By contrast, Zhan et al. proposed a blind watermarking technique based on the fluctuation values of the root mean square curvature. The watermark was embedded by modulating the mean normalized fluctuation values of the root mean square curvature. The experimental results showed that the algorithm had good imperceptibility and provided good resistance against vertex rearrangement, rotation, translation, uniform scaling, noise, smoothing, quantization, and simplification [5]. Singh et al. introduced a non-blind watermark detection scheme based on the computed distance between each vertex and the center of gravity of the 3D object. The watermark was divided into patches with size of 128 and hashed using MD5 using the sum of vertex normal in one group of vertices. Then, the authors performed Exclusive or (XOR) between the result from applying hash function and the watermark bits. The result was embedded in vertices of the mesh model. The proposed algorithm was resistant against translation, rotation, and uniform scaling attacks [6]. Amar et al. presented blind detection schema. The watermark bits were embedded by decreasing the Euclidian distance from the center mass of the faces to the mass center of the mesh. The robustness of the proposed algorithm was improved by applying turbo encoding to correct the errors caused by transmission or distribution. The performance evaluation showed good results in terms of imperceptibility and robustness [7].

To obtain a 3D object watermarking algorithm that provides results with high imperceptibility and robustness, researchers have often utilized artificial intelligence techniques in selecting appropriate places in a 3D object to embed the watermark. El Zein et al. introduced a non-blind watermarking algorithm using K-mean clustering technique for clustering the 3D mesh vertices to find the best positions to embed the watermark. The clustering operation was performed on the basis of the computed angles between the surface normal and average normal, which were inputted as feature vectors. The algorithm was robust against different geometric attacks, such as cropping, noise addition, and smoothing. The algorithm also introduced high results of imperceptibility [8]. Laftah and Luma proposed a blind 3D mesh watermarking algorithm based on subdividing the faces of the mesh to remove noise and increase payload. The watermarking bits were encrypted by applied XOR with normal direction of every vertex. K-mean clustering was applied to the vertices, and the cluster that had the largest silhouette value was selected for embedding. The algorithm showed good imperceptibility and robustness against different types of attack [9]. El Zein et al. also introduced a non-blind watermarking algorithm in 3D mesh media. They used K-mean technique to cluster the mesh vertices. Feature vector was constructed by computing the angles between the surface normal and average normal for vertices that had a valance of 6. The performance evaluation showed high imperceptibility and good robustness against smoothing, noise, and cropping [10]. Motwaniet al. presented a non-blind algorithm by using a linear support vector machine classifier (SVM) to classify vertices whether they were suitable or candidates for watermark embedding. Feature vectors were derived from the curvature estimates of a one-ring neighborhood of vertices and used to train the SVM. The robustness of the proposed algorithm showed acceptable resistance against smoothing, noise, and cropping attacks [11].

In this study, we present an efficient digital 3D model watermarking algorithm based on altering the geometrical properties of a 3D model. The embedding algorithm is performed by classifying the points of the model to appropriate and inappropriate regions to construct an optimal domain that is minimally affected by embedding. The proposed algorithm uses a secret text as watermark, which has been embedded carefully in selected vertices depending on the values of their MC, as explained mathematically by Abbena et al. [13] and illustrated in Sect. 3. The proposed algorithm provides blind extraction that does not require prior knowledge about the original model during watermark extraction. The proposed algorithm is evaluated through evaluation measurements, and the effects of various types of attack are investigated. The results indicate that the proposed algorithm is effective and has high level of resistance against numerous types of attack.

2 Implementation of 3D Model Watermarking

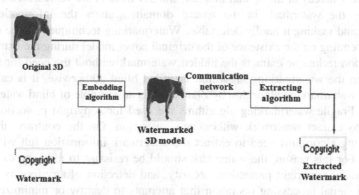

Fig. 2. General framework of the 3D model watermarking system

The general framework of the 3D model watermarking system constructed from the embedding and extracting algorithm is shown in Fig. 2.

At the sender side, the owner of the original 3D model needs an embedding algorithm to hide a small label called "watermark," which is either a small image, text, or group of bits, to protect the original data from being illegal copied and distributed and detect any tampering from an attackernd send the watermarked data as a result through the communication network. In this step, imperceptibility is measured to ensure no noticeable difference between the original and watermarked models exist. At the receiver side, the extraction algorithm is used to extract watermark information from modified regions and compares with the original one to identify the sender. Subsequently, the robustness is calculated to show that the watermark is still present despite of its vulnerability from previous several types of attack. In this study, the proposed watermarking algorithm is blind. Thus, the algorithm does not need the original model during extraction, and the embedded regions are the vertices that have positive MC values.

The performance evaluation to any watermark system usually needs to satisfy certain requirements as follows [12].

1. Imperceptibility: Imperceptibility implies that the hidden watermark is completely invisible. The human visual system (HVS) cannot distinguish between the data that

contain the watermark from the original. This process is performed by hiding the watermark, such that it does not minimize or distort the quality of the original data.

2. Robustness indicates that the hidden sensitive watermark is still alive and can be fully extracted after applying numerous types of attack.

3. Payload: Payload implies that the number of bits can be hidden in the original data without the effect on the perceptual quality of these data. Generally, 60–70 bits of information should be embedded in the original data.

Figure 3 summarizes different categories of 3D model watermarking techniques. In the frequency-domain watermarking, the 3D models are represented in terms of their frequencies, whereas in the spatial domain, the 3D models are represented by pixels. Embedding the watermark in the spatial domain scatters the information to be embedded and making it hardly detectable. Watermarking techniques can be classified further depending on the existence of the original cover model during the extraction. If the extraction technique extracts the hidden watermark without the need for the cover model, then the watermarking algorithm is called blind. Otherwise, it is called non-blind. The watermarking algorithm by Zhan et al. is an example of blind watermarking algorithm. Fragile watermarking algorithms are used for copyright protection and do not need to extract watermark without any distortion. On the contrary, the robust watermarking algorithms need to extract the watermark information full without any distortion. For this reason, the watermark should be resistant to many types of attack and is used for copyright protection, security, and detection of any forgery [16]. An attack is a signal processing operation that attempts to destroy or minimize the perceptual quality of a hidden watermark. The attacks can be divided into removal

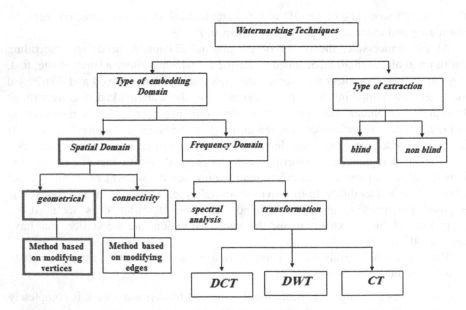

Fig. 3. 3D model watermarking techniques

(malicious) attacks, such as noise and smoothing; and geometrical (affine) attacks, such as translation, rotation, and scaling attacks [17].

3   Proposed 3D Model Watermarking Algorithm

This section explains the mathematical background of the MC for points in any surface.

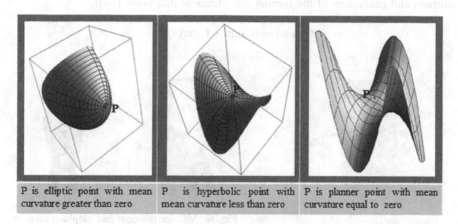

| P is elliptic point with mean curvature greater than zero | P is hyperbolic point with mean curvature less than zero | P is planner point with mean curvature equal to zero |

Fig. 4. Classification points based on MC

MC is an important measurement of the curved surface f(u, v). It can be used to classify the surface points. Figure 4 shows the classification of a point in a surface based on its MC value.

MC is calculated for every point in the surface by taking one-half of the sum of the principal curvatures at that point. The MC is calculated as follows:

$$MC = (k1 + k2)/2, \tag{1}$$

where k1 and k2 are two vectors called the principle curvatures that identify the local shape at a given point. To compute for MC, we need to find the 1st and 2nd fundamental forms of the surface, that is,

$$1\text{stfundamental form} = Edu^2 + 2Fdudv + Gdv^2, \tag{2}$$

$$2\text{ndfundamental form} = Ldu^2 + 2Mdudv + Ndv^2. \tag{3}$$

If $X \to U^3$ is a regular patch, then the MCis computed as

$$H = (LG - 2MF + NE)/(2(EG - F^2)), \tag{4}$$

where E, F, and G are the coefficients of the 1st fundamental form; and L, F, and G are the coefficients of the 2nd fundamental form. The sign of the MC depends on the choice of the normal curvatures [13]. Normal curvatures at any point in the surface are the curvature of the curve on the surface in plane containing the tangent vector at this point, as shown in Fig. 5. The principal curvature of a surface at a point depends on the minimum and maximum of the normal curvatures at that point [14].

$$Cmean = (Cmin + Cmax)/2, \tag{5}$$

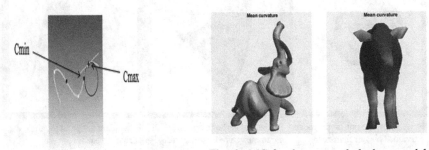

Fig. 6. MC for the cow and elephant models

Fig. 5. Principal curvatures of a surface

where Cmean is the MC of any point on the surface depending on its normal curvatures. Cmin and Cmax are the minimum and maximum curvatures, respectively. Figure 6 shows the MCs for the cow and elephant test samples.

3.1   Watermark Embedding Algorithm

The use of watermarking with 3D models depends on spatial domain techniques, which provide better robustness than frequency domain techniques. After studying the geometrical properties of the 3D model, we found that the MC of points in the 3D model can be used as criteria for embedding.

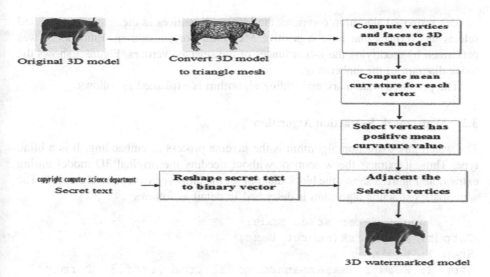

Fig. 7. Watermark embedding algorithm

Input: 3D model stored in .off file format and secret text as a watermark.
Output: 3D watermarked model.
Step 1: Begin
Step 2: Read 3D model from .off file format.
Step 3: Convert 3D model to triangle mesh.
Step 4: Read vertices and faces from the 3D model and store them in faces (F) andvertices (V) matrices, respectively.
Step 5: Read secret text and reshape it to binary vector W= [w1, ...,wn], where n is the length of the secret text.
Step 6: Compute the MC for all vertices depending on Equation 4.The value of MC is positive, zero, or negative.
Step 7: Select a vertex with a positive MC value and convert it into binary
Step 8: Embed one bit from w[i] by modifyingthe z-coordinate for the selected vertices depending on the following equations:
$$V'(x, y, z) = V(x, y, z) + W[i], (6)$$
where V' isthe watermarked vertices; V isthe original selected vertices, and w is thewatermark vector.
Step 9: Repeat Steps 7 and 8 until all secret dataare embedded.
Step 10: Save and display the watermarked model(M).
Step 11: End

Our proposed algorithm computes the MC of all vertices in the original model and selects the vertices that have a positive MC value for embedding. Embedding was performed by modifying the z-coordinates of the selected vertices. Figure 7 shows the block diagram of the embedding.

The proposed watermark embedding algorithm is explained as follows.

3.2   Watermark Extraction Algorithm

The watermark extraction algorithm is the inverse process to embedding. It is a blind type. Thus, it extracts the watermark without needing the original 3D model during extraction. Figure 8 shows the block diagram of the proposed extraction algorithm. The watermark extraction algorithm is described in detail as follows

```
     Input: Watermarked model
Output: Watermark(secret text)
Step 1: Begin
Step 2: Read 3D watermarked model from .offfile format
(M).
Step 3: Convert the 3D watermarked modelinto a triangular
mesh.
Step 4: Read vertices and faces from the mesh and store
them in (V) and (E) metrics.
Step 5: Compute the MC for each vertex using Equation 4
Step 6: Separate the mean values, such that the value is
either positive, negative, or zero.
Step 7: Select the vertex position with a positive mean
value for extraction.
Step 8: I=1
Step 9: Extract one bit from the selected vertex and
store them in the watermark vector.
Step 10: I=I+1
Step 11: Repeat Steps 9 and 10 until I> length (secret
text).
Step 10: Convert the watermark vector from binary to
character
Step 11: Display watermark.
Step 12: End
```

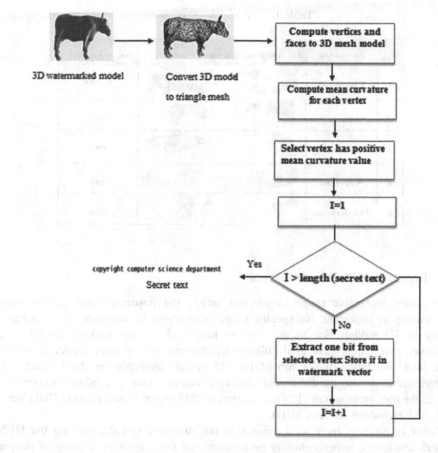

Fig. 8. Watermark extraction algorithm

4 Experimental Results and Discussion

Figure 9. The performance of watermarking algorithms is usually evaluated with respect to two properties, namely, imperceptibility and robustness. We use five 3D models of type .off file format downloaded for free from the Stanford University database as test samples [3]. The information that describes each model is explained in Table 1. The models vary in number of vertices and the nature of their surface. As a watermark, we use two sizes of texts. We start embedding text from 66 bits to 151 bits.

Table 1. Test sample description

No.	3D model	Original and Mesh Models	No. of vertices
1	Elephant		24955
2	Cow		726
3	Camel		586
4	Nefertiti		299
5	Mushroom		226

4.1 Imperceptibility Measurements

Determining acceptable range values that satisfy the imperceptibility performance requirements is important. No specific range is available to determine the imperceptibility of 3D models due to the discrete nature of vertices in such models [4]. Moreover, researchers have used different file formats for the same model given the more than 140 types of file format for 3D models available on the Internet [2]. Researchers have also used different objective measurements to evaluate invisibility, such as the root mean square (RMS), maximum RMS signal-to-noise ratio (SNR), peak SNR, and Hausdorff distance (HD).

Other researchers have used subjective measurement that depends on the HVS through testing the imperceptibility performance of the system by a group of people whose stated their score and calculated the mean of their scores[4].

In the present work, the imperceptibility of the proposed algorithm is verified by calculating the SNR between the vertices of the original 3D model and the vertices of its watermarked version, as shown as follows [8]:

$$\text{SNR} = \frac{\sum_{i=1}^{N}\left(x_i^2 + y_i^2 + z_i^2\right)}{\sum_{i=1}^{N}\left[\left(x_i - x_i'\right)^2 + \left(y_i - y_i'\right)^2 + \left(z_i - z_i'\right)^2\right]}, \tag{7}$$

where N is the number of vertices; x_i, y_i, and z_i are the coordinates of the vertices in the original model; and x_i', y_i', and z_i' are the new watermarked coordinates.

The HD is also used to measure imperceptibility by calculating the error between two sets of vertices before and after embedding. HD is defined as follows. For every point of model A, we find the minimum distance to any point b of model B. Then, HD is the maximum distance among all points to point a, as calculated as follows [9]:

$$HD(A, B) = \max_{a \in A}\left\{\min_{b \in B}\{d(a, b)\}\right\}. \tag{8}$$

A high SNR value implies a stronger signal power in the original 3D model than the noise level in the watermarked 3D model. Meanwhile, the efficient watermarking algorithm provides minimum HD values by reducing the error between the original 3D model and its watermark version. Table 2 shows the experimental results of imperceptibility measurements between the original 3D model before and after hiding the secret text.

Table 2. Imperceptibility measurements

No.	Original models	Embedding watermark 66 bits		Embedding watermark 151 bits	
		HD	SNR	HD	SNR
1		0.177	138.39	0.177	133.28
2		0.002	77.92	0.002	73.65
3		0.178	101.32	0.178	99.44
4		0.0039	89.36	0.0039	122.44
5		0.002	100.99	0.002	115.04

From the results shown in Table 2, the SNR values are affected by the length of the secret text. When the payload increases, the SNR value also increases for the elephant, cow, and camel models. Meanwhile, the SNR values for the other two models increase with the size of the hidden message due the differences in their natural surface and distribution of changeable area.

4.2 Robustness Measurements

The correlation Factor (CF) is used to measure the similarity between two sequences of bits. The values of correlation are between 1 and −1. In the 3D model watermark domain, correlation is calculated between the original watermark and the watermark extracted from the attacked 3D watermarked model, as shown as follows [15]:

$$CF = \frac{\sum_{i=1}^{N-1}\left(w_i' - \overline{w_i'}\right)\left(w_i' - \bar{w}\right)}{\sqrt{\sum_{i=1}^{N-1}\left(w_i' - \overline{w_i'}\right)^2 \cdot \sum_{i=1}^{N-1}\left(w_i' - \bar{w}\right)^2}}, \tag{9}$$

where $\overline{w_i'}$ and $\bar{w}$ indicate the average of the extracted watermark bit sequence w_i' and the average of the inserted watermark bit sequence, respectively. Generally, CF = 1 indicates that the hidden message is extracted completely without any distortion and CF, where in approximately 0.75 or above is considered a good value.

In the next two sections, we show the experimental results after applying numerous types of geometrical and removable attacks to the watermarked model that containin visible secret text with length of 151 bits.

Effect of Watermark Geometrical Attack

Geometric attacks do not remove the embedded watermark but intend to distort it. The receiver can reconstruct the hidden watermark when perfect extraction algorithm is used. The geometric attacks that we use to evaluate our algorithms are as follows.

Rotation. Rotation is the process of rotating a 3D object to a specific angle in different directions around a specific coordinate. Figure 10 shows the camel model after rotating at 70° around the z-coordinate.

Scaling. Scaling is a process of changing the size of a 3D object uniformly by increasing and decreasing the objects by a fixed scale factor in all directions along a specific axis or plane. The experiment shows that the x-coordinate is more affected by scaling than others. Figure 11 shows the camel model after scaling of approximately 0.5 in the x-coordinate.

Translation. is the process of moving every point of a 3D object by the same distance toward a specific direction. Specifically, translation adds a constant vector P to every vertex V in the 3D mesh. Thus, the translation is defined as

$$V\,new = V\,old + P. \tag{10}$$

Figure 12 shows the camel model after being translated by −5.

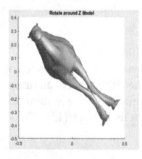

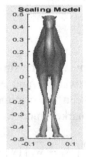

Fig. 9. Rotated model   **Fig. 10.** Scaled model   **Fig. 11.** Translate model

Table 3 shows the experimental results of the CF measurement between the watermark extracted from the attacked watermarked models after applying geometrical attacks, including rotation, scaling, and translation, and its original version.

Table 3. CF measurements after applying geometrical attacks

3D model	Correlation								
	Rotation			Scaling			Translation		
	70	120	240	0.5	0.9	2	−5	2	5
Elephant	1	1	1	1	1	1	1	1	1
Cow	1	1	1	1	1	1	1	1	1
Camel	1	1	1	1	1	1	1	1	1
Nefertiti	1	1	1	1	1	1	1	1	1
Mushroom	1	1	1	1	1	1	1	1	1

Effect of Watermark Removal Attack

Removable attacks attempt to destroy or distort the hidden watermark in 3D models by altering the important components in the construction of these models. The removable attacks that we use to evaluate our algorithms are as follows.

Smoothing. Refers the process of enhancing the model quality while the model topology remains unchanged by shifting some model vertices. It divides each face into N^2 small faces, where N is an integer number > 1. Figure 13 shows the camel model after applying smoothing when N = 2.

Noise Affects the Watermark. All vertices in the 3D watermarked model are affected after adding the same level of noise. Figure 13 shows the camel model after adding a noise level of 0.1.

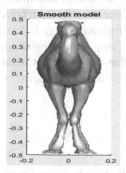

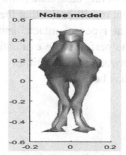

Fig. 12. Smoothed model   **Fig. 13.** Noised model

Table 4 displays the experimental results of the CF measurement between the watermark extracted from the attacked watermarked models and the original watermark after applying smoothing and noise attacks.

Table 4. CF measurements after applying removable attacks

3D model	Correlation					
	Smoothing			Noise		
	2	3	4	0.001	0.0001	0.0001
Elephant	1	1	1	0.17	0.90	1
Cow	1	1	1	0.36	0.80	1
Camel	1	1	1	0.31	0.72	0.92
Nefertiti	1	1	1	0.20	0.82	1
Mushroom	1	1	1	0.58	0.77	0.94

The experimental results show that the proposed algorithm introduces high robustness and that the hidden text is extracted completely before applying attacks and after applying a certain level of geometrical and smoothing attacks. Meanwhile, increase in the noise reduces the perceptual quality of the hidden text. The proposed algorithm has achieved significant improvement in robustness compared with the method proposed by Amar et al. [7], which achieved correlation results equal to 1,1,0.73, and 0.95 after applying cropping subdivision, smoothing, and noise attacks to a cow model, respectively.

5   Conclusion

In this work, a new robust and blind watermarking method is proposed to 3D models for copyright protection purposes. The watermark has been hidden in the vertices of the 3D model based on their MC values. The proposed algorithm minimizes the surface distortion by selecting candidate points, which introduce the minimum error. Moreover, the experimental results prove that the proposed algorithm has high level of robustness against different types of attack, such as rotation, cropping, translation, noise, and smoothing. The performance evaluation indicates that the proposed algorithm solves the tradeoff between imperceptibility and robustness. Thus, the balance between these requirements has been set.

References

1. Fabio, G., Massimiliano, C., Sumanta, P., Marco Di, B.: Introduction to Computer Graphics, vol. 55. Addison-Wesley, Reading (1994)
2. McHenry, K., Peter, B.: An overview of 3D data content, file formats and viewers. Natl. Cent. Supercomput. Appl. **1205**, 22 (2008)
3. The Stanford 3D scanning repository. http://www-graphics.stanford.edu/data/3dscanrep
4. Garg, H., Suneeta, A.: A secure image based watermarking for 3D polygon mesh. Sci. Technol. **16**(4), 287–303 (2013)
5. Zhan, Y., Li, Y., Wang, X., Qian, Y.: A blind watermarking algorithm for 3D mesh models based on vertex curvature. J. Zhejiang Univ. Sci. C **15**(5), 351–362 (2014)

6. Singh, L.K., Chaudhry, D., Varshney, G.: A novel approach of 3d object watermarking algorithm using vertex normal. Int. J. Comput. Appl. **60**(5) (2012)
7. Amar, Y., Trabelsi, I., Dey, N., Bouhlel, M.: Euclidean distance distortion based robust and blind mesh watermarking. Int. J. Interact. Multimed. Artif. Intell. **4**(2), 46–51 (2016)
8. El Zein, O.M., El Bakrawy, L., Ghali, N.: A non-blind robust watermarking approach for 3D mesh models. J. Theor. Appl. Inf. Technol. **83**(3), 353 (2016)
9. Laftah, M., Luma, F.J.: Data hiding in 3D model using artificial techniques. Int. J. Sci. Technol. **143**(3077), 1–9 (2016)
10. El Zein, O.M., El Bakrawy, L., Ghali, N.: A non-blind robust watermarking approach for 3D mesh models. J. Theor. Appl. Inf. Technol. **83**(3), 353 (2016)
11. Motwani, R., Motwani, M., Bryant, B., Harris Jr, F., Agarwal, A.: Watermark embedder optimization for 3D mesh objects using classification based approach. In: International Conference on Signal Acquisition and Processing, 2010. ICSAP'10. IEEE (2010)
12. Dugelay, J., Baskurt, A., Daoudi, M.: 3D Object Processing: Compression, Indexing and Watermarking. Wiley, New York (2008)
13. Abbena, E., Salamon, S., Gray, A.: Modern Differential Geometry of Curves and Surfaces with Mathematica, 2nd edn, pp. 111–115. CRC Press, Boca Raton (1997)
14. Cohen-Steiner, D., Morvan, J.: Restricted delaunay triangulations and normal cycle. In: Proceedings of the Nineteenth Annual Symposium on Computational Geometry. ACM (2003)
15. Wang, K., Lavoue, G., Denis, F., Baskurt, A.: A comprehensive survey on three-dimensional mesh watermarking. IEEE Trans. Multimed. **10**(8), 1513–1527 (2008)
16. Luo, M.: Robust and blind 3D watermarking. Dissertation. University of York (2010)
17. Rolland-Névière, X.: Robust 3D watermarking. Dissertation. Université Nice Sophia Antipolis (2014)

Machine Learning

Multi-objective Feature Selection: Hybrid of Salp Swarm and Simulated Annealing Approach

Muntadher Khamees[1]([⊠]) [iD], Abbas Albakry[2] [iD], and Khalid Shaker[3] [iD]

[1] Computer Science, College of Science, Diyala University,
Baquba, Diyala, Iraq
alkarawis@gmail.com
[2] University of Information Technology and Communications, Baghdad, Iraq
abbasm.albakry@uoitc.edu.iq
[3] College of Computer Sciences and Information Technology,
University of Anbar, Anbar, Iraq
Khalidalhity@gmail.com

Abstract. Met-heuristics are becoming increasingly popular in solving real world problems. Modern meta-heuristics leading to a new branch of optimization, called meta-heuristic optimization. These applied to all areas of data mining, planning and scheduling, design, machine intelligence, and features selection (FS). FS is used to remove noise from data and dimensionality reduction; these properties could give rise to simplicity of rules, speed of learning, predictive accuracy and visualizes the data. Salp Swarm (SSA) is a recent meta-heuristic optimization method that mimics the innate behaviour of the Salp swarm chain. In this study, SSA is hybridised with a simulated annealing (SA). SA is employed as internal functions to improve the exploitation ability that utilizes to accept a worse quality solution than the current one. The performance of suggested approach is evaluated on 16 datasets including two high dimensional from UCI repository and compared with the native (SSA) and other (FS) approaches include ALO and PSO, the experimental results clearly proved the adequacy of the proposed approach to search the features space for optimal features. SSA-SA gave excellent performance as a multi objective optimisation where achieved two contradictory goals, maximal accuracy of a classification with minimal size of features on all used datasets.

Keywords: Meta-heuristics · Wrapper model · Salp chain
Feature subset selection · Simulated annealing technique

1 Introduction

Knowledge discovery in databases (KDD) has been produced for classifying datasets in order to obtaining beneficial information from a large scale data. KDD includes four major processes namely, warehousing of data, preprocessing, data mining and post processing [1, 2]. It is of interest to researchers in knowledge acquisition, machine

© Springer Nature Switzerland AG 2018
S. O. Al-mamory et al. (Eds.): NTICT 2018, CCIS 938, pp. 129–142, 2018.
https://doi.org/10.1007/978-3-030-01653-1_8

learning, pattern recognition, databases, artificial intelligence, data visualization, and statistics. Data mining techniques is an essential part in Knowledge discovery. In data mining, the quality of data or how to find useful information is an important issue that may effect of classification performance and prediction model. Data mining attempts to gather data and analyze it to get useful knowledge, it modes include regression, classifying, clustering or grouping, deviation detection, dependency modeling, change detection and summarization [3, 4].

Dimensionality problem make the learning job more complex, and computationally expensive. It occurs when the dataset has the following properties, very high number of samples or features or both. The reduction methods can be applied to improve the predictive accuracy, simplicity of rules and learning speed. The irrelevant and redundant dimensions in the dataset are discarded to improved discriminant power [5]. Features are also known as attributes, properties, dimensions and characteristics. The growing demand of taking advantages of high-dimensionality data stored form a new challenge for data mining. FS is one of the solutions that employed to create a predictive that minimizes the prediction errors of classifiers through select important features and discarding irrelevant and noisy features in original data set [6, 7].

FS methods usually classified into three broad categories are: wrapper, filter, and embedded models [8]. Filter-based methods are rankers; the features are evaluated according to measures directly from the data without predictors and using rank ordering of the features. The wrapper-based approach uses the learning methods as a classifier (as part of the fitness function) to evaluate the usefulness of subset features and subsequently obtaining the better performance of prediction. Embedded methods work with linear classifiers such as SVM, are embedded in the algorithm as expanded functionality. Feature selection (FS) method is a search problem, Search methods can be classified as follows: exhaustive, heuristic, probabilistic, and automatic hybrid search algorithm. Heuristic strategy is much faster, time consuming and only searches a particular track to obtaining the best neighbour [7]. Heuristic usually applied to real-world problems, and to a very wide range of computer scientist [9]. Furthermore, heuristics are appropriate to deal with other aspects of huge Data, such as, variety, and velocity. Heuristic-based search design shows two criteria are exploitation to determine the best neighbour (intensification) and exploration (diversification) in the search area [10].

Various heuristics technicality has been utilized to deal with attributes selection and a survey can be found in [11]. Genetic algorithm (GA) is the most inspected metaheuristics [12]. Both single and population-based heuristics have been suggested. In single-based met heuristics, simulated annealing [13], Hill climbing [14], Tabu Search [15], harmony search [16], there are two major drawbacks of Hill climbing, is very ticklish to the initial solution and it oftentimes falling in local optima. In population-based, all the approaches have been used from classical mimetic algorithm, genetic programming [17], particle swarm optimization [18], to bat algorithm [19], ACO [20].

In November 2004, Lee and Oh introduced the first hybrid FS approach. Recently, hybrid heuristics techniques showed high-performance in solving data hard combinatorial optimisation problems. Aggregation approaches such as hybrids between GA with PSO [21], and ACO with GA [22] have also been proposed, furthermore, there are several methods that employed local-search algorithms as an internal operator to balance between diversification and intensification [23]. Moreover, in [24], proposed hybridisation between GA and PSO which used the support vector machine (SVM) as a classifier? For recent hybrid techniques we refer to [25].

There is no heuristic based model is able to solve all FS problems. Notwithstanding, improvements can be made to exist search space algorithm to obtain high performance regions of search area. This motivation underlies most our experiments to produce a predictive model based on hybridized between a global search with local search algorithm, this paper proposed a new hybrid method named (SSA-SA) to enhance the exploitation ability of the native SSA.

The rest sections is organised as following: A brief background of the fundamental SSA and SA algorithms are described in Sect. 2. Section 3 provided the detailed description of the main inspiration of this paper and proposed approach, the datasets, parameters and experimental results are represented in Sect. 4. Eventually, Sect. 5 concludes the work, major achievement and suggests future research studies.

2 Salp Swarm Algorithm

Salp is marine animal belonging to the subfamilies in Salpidae. It's gelatinous, transparent and tubular -shaped body. Their texture is comparable to a jellyfish. To propel themselves along, Salp pumps water over its body. In deep cold Ocean, Salps are aggregating while swimming and feeding into a chain shaped group called blastozooids [26]. This behaviour achieves highlighted method to the most superb locomotion by using swift harmonise changes and foraging.

In December 2017, [27] proposed a swarm algorithm based on Salps behaviour to solve optimisation problems. Originally, the population of the Salps chain is classified into two portions: leader at the front and the remaining Salps at rearward of the chain that called followers. The followers trace back the leader Salps in the direction of movement. The initial positions are represented as 2D matrix called x. FS is the goal food source. Only the leader location modifies with regard to the feed place, see Eq. (1).

$$y_j^1 = \begin{cases} FS_j + p_1\big((ub_j - lb_j)p_2 + lb_1\big)p_3 \geq 0 \\ FS_j - p_1\big((ub_j - lb_j)p_2 + lb_1\big)p_3 < 0 \end{cases} \tag{1}$$

where yj1 are leaders location in the jth, FSj shows the food source location. ubj is upper & lbj is lower boundary. p1 is a random parameter utilized to balancing among exploitation & exploration in the search space, p1 can be calculated as follows Eq. (2):

$$p_1 = 2e^{-\left(\frac{4l}{L}\right)^2} \tag{2}$$

where l shows the current & L is the max iterations. p_2, p_3 are random numbers among [0, 1], they determine the next location of chain if towards positive or negative infinity in ith dimension. Followers update position based on Newton's law of motion as following Eq. (3):

$$y_j^i = \frac{1}{2} a t^2 + v_0 t, \tag{3}$$

Where i is ≥ 2, y_j^i show the followers location in jth dimension, t is time &v_0 indicated to the initial velocity, where a is formulated as following Eq. (4):

$$a = \frac{v_{final}}{v_0} where = \frac{y - y_0}{t}, \tag{4}$$

where a is equal to v final that divided by v_0 & $v = y - y_0$ divided by t. In optimisation, the difference between iterations is equal to one. Equation (5) can be calculated as following:

$$y_j^i = \frac{1}{2} \left(y_j^i + y_j^{i-1} \right), \tag{5}$$

where $v_0 = 0$, and y_j^i indicate the location of followers salp chain, where $i \geq 2$ in ith dimension. Figure 1 indicates the pseudo code of SSA. By Eqs. (1) & (5), the Salp chains can be mimicked [27].

3 Simulated Annealing

In [28], Kirkpatrick et al. proposed a single-based solution method called simulated annealing. SA is a non-traditional technique for obtaining a best neighboring solution to an optimisation problem that simulates the process of the solid annealing principle [29]. The fundamental property of meta-heuristics is accepting some worse quality neighbours. These allows for a more extensive search to avoid falling in local optima and find better solutions. The worse solutions are accepted according to a certain probability of a physical system specified by the probability of Boltzmann distribution probability $P = e^{-\theta/T}$, $\theta = f(Sol*) - f(Sol)$ where f indicate the fitness function, Sol is the Current best neighbour and $Sol*$ is a new neighbour solution. Furthermore, T is the absolute temperature which begins with a high temperature, then slowly decreases and gradually orderly during the search space process, until arrive to an equilibrium state. In this study, the primary temperature value set to 0.1. As adopted in [30], the cooling schedule is formulated as $T = T * 0.93$. SA can be performed as the Fig. 2.

4 Proposed Hybrid SSA with SA Approach

FS problems considered as binary search problems. A binary Salp swarm algorithm adjusts the continuous Salp to deal with features selection in bilateral bound of dimension. Each solution is symbolized as a binary vector with one dimension. The vector size is equivalent to the number of the features in original dataset. Inside the vector, each cell is designated by one or zero. 1 value indicate that the selected features otherwise indicate to neglected features.

```
Initialize the salp populationyᵢ(i=1, 2 …n) consid-
ering ub, lb
While (end condition is not satisfied)
        Calculate the fitness of each search agent
(salp)
        FS=the best search agent
        Update cᵢby equation (2)
For each salp (yi)
If (i==1)
            Update the position of the leading
Salp    by eq. (1)
else
            Update the position of the follower
Salp by eq. (5)
end
end
        Amend the salp based on the upper and lower
bounds of variables
end
Output FS
```

Fig. 1. Pseudocode of native SSA.

In this paper, suggested fitness function is utilized to evaluate each agent in the search space of SSA-SA that depends on KNN as classifier, where (K = 5). KNN is one of the simplest nonparametric predictor methods of supervised learning techniques, which based on error and trial [31], a new unknown record is a classifying based on the distance between the new record and the training record set [32]. The proposed fitness function can be calculated as the following Eq. (6):

$$fitness = \alpha * Er(D) + \beta \frac{|L|}{|T|})$$

(6)

where Er(D) is classifier error rate, |L| indicated the features length, |T| is total number of attributes, the α & β are constant utilized to control on subset length and performance of the classification, α is an integer value belong to range between [0,1], β is 1−α adopted by [33, 34]. The inbred behaviour of the Salp chain is a recent mimic method that utilized to a very wide range of optimisation problems in computer scientist to obtaining the best neighbour. This approach introduced hybridization between global and local search algorithms, were SA is embedded within Salp swarm algorithm to enhance the exploitation space by searching the most optimistic areas located by SSA algorithm. SA algorithm employed after applying SSA algorithm to enhance that final solution through replacing the original SSA solution by the SA enhanced one. Moreover, genetic agents, as mutation and crossover of the suitable are employed as internal functions in SA. The mutation utilized to create a new Solution with (pm = 0.05), while the Crossover takes that worst solution with the best current solution as inputs to find the updated to the next position, as showed in the Fig. 2.

```
Initialization Phase
% Set maximum number of iteration, max_Iter;
% Set initial solution, Sol;
% Set best solution, BestSol ← Sol;
% Set initial temperature T;
Improvement phase
Iter ←0;     T← T0;
%Compute fitness value for initial solution (Sol);
Do while (Iter < max_Iter)
        Sol*←Mutate (Sol); % create a new neigh-
borhood of Sol (Sol*);
        If (fitness (Sol*) < fitness (BestSol)
              BestSol ← Sol*;
              Sol ← Sol*;
else % Accepting a worst solution
              % Calculate the acceptance probabil-
ity of Sol*,
              θ = fitness (Sol*) - fitness (Sol);
              P=exp (-θ /T);
              % Generate a random number in [0, 1];
              If (Rand <= P)
                    Sol+ = Crossover (Sol, Sol*);
                    Sol ← Sol+;
                    T← T * 0.93; %update temperature
        end if
        end if    Iter ++;
        end do
Output BestSol
```

Fig. 2. Pseudocode of enhanced SA.

5 Empirical Result and Discussion

5.1 Dataset and Parameters

The empirical results were evaluated on 16 datasets including two high dimensional from UCI repository of California university [35]. All utilized dataset details are represented in Table 1, the suggested hybrid approach is a wrapper-based technique. Each individual in population is symbolized as index binary vector for the attributes in dataset. In the basic SSA-SA, we hold only the most optimistic solution and its fitness, which achieved maximum accuracy with the minimal number of features. In this work, the maximal iterations number is 80 and the search agent number is 5. Furthermore, all results were calculated with an average of 20 runs using Matlab framework on Intel Corei7 machine, 2.67G CPU and 4.00 G of RAM with 64-bit Operating system.

Table 1. Dataset representation.

DataSet	No. of features	No. of instances	Area of dataset
Breastcancer	9	699	Life
Exactly	13	1000	N/A
Exactly2	13	1000	N/A
Breast_EW	30	699	Life
Congress_EW	16	535	Social
Heart_EW	13	270	Life
Ionosphere_EW	34	351	Physical
Krvskp_EW	36	3196	Agricultural
Lymphography	18	148	Life
M-of-n	13	1000	Rules
Sonar_EW	60	208	Physical
Spect_EW	22	267	Life
Tic-tac-toe	9	958	Game
Waveform_EW	40	5000	Physical
Wine_EW	13	178	Physical
Zoo	16	101	Life

5.2   Evaluation Criteria

Each dataset is separated into three different sections; the first section is utilized to train classifiers throughout optimisation process. The Second section is used for evaluation the classifiers performance through the optimisation time and the third section is to provide an equitable assessing of a conclusive approach.

To compare the different FS approaches with the suggested hybrid approach, we utilized the different following indicators, statistical best, worst; mean fitness, standard deviation and average selection size, as the following mathematical Equations:

$$Bestfitness = min_i^M = 1g_*^i,$$
(7)

$$Worstfitness = miniM = 1g * i$$
(8)

$$Meanfitness = \frac{1}{M} \sum_{i=1}^{M} g_*^i$$
(9)

$$Standarddeviation = \sqrt{\frac{1}{M-1} \sum (g_*^i - mean)^2},$$
(10)

$$Average selected size = \frac{1}{M} \sum_{i=1}^{M} \frac{size(g_*^i)}{D}, \tag{11}$$

5.3   Results and Discussion

In this work, SSA is hybridised with a SA. SA is employed as internal functions to improve the exploitation ability that utilizes to accept a worse quality solution. Experimental Results of proposed approach was compared with the native salp swarm algorithm and other features selection algorithms, including ALO and PSO, on the basis of the valuation criteria's that mentioned in the previous section.

The performance of SSA-SA was compared with that of Salp swarm on the basis of two objectives, namely, classification accuracy and the size of selected features. For multi objective optimisation for wrapper based models, as for hybrid approaches, the objectives are: maximizing of the classifier quality and on the other hand, minimizing of the features selected size [15]. As shown in Tables 2, 3, we can evidently remark that the proposed hybrid method is considerably outperforms that of the native SSA. The empirical results of SSA-SA indicated the most excellent performance as a multi-objective optimisation where achieved two contradictory targeted, minimal size of features with maximal classification accuracy on all used datasets.

Table 2. Comparison between SSASA and other optimisers in terms of classification accuracy.

Dataset	Full	PSO	ALO	SSA	SSA-SA
Breastcancer	0.94	0.9567	0.9608	0.9611	0.9714
Exactly	0.67	0.7431	0.7131	0.7104	0.9999
Exactly_2	0.74	0.6778	0.7003	0.6976	0.7547
Breast_EW	0.96	0.9295	0.9370	0.9332	0.9654
Congress_EW	0.92	0.9257	0.9362	0.9303	0.9729
Heart_EW	0.82	0.7859	0.7904	0.7863	0.8508
Ionosphere_EW	0.87	0.8634	0.8849	0.8778	0.9332
Krvskp_EW	0.92	0.9102	0.9115	0.9031	0.9708
Lymphography	0.68	0.7566	0.7922	0.7830	0.8850
M-of-n	0.85	0.8397	0.8345	0.8289	0.9992
Sonar_EW	0.62	0.8486	0.8611	0.8601	0.9341
Spect_EW	0.83	0.7817	0.7847	0.7884	0.8563
Tic-tac-toe	0.72	0.7578	0.7641	0.7634	0.7943
Waveform_EW	0.77	0.7142	0.7115	0.7096	0.7460
Wine_EW	0.93	0.9461	0.9590	0.9551	0.9910
Zoo	0.79	0.9333	0.9499	0.9441	0.9794

Table 3. Comparison between SSA-SA and other optimizers in terms of Average selected size.

Dataset	PSO	ALO	SSA	SSA-SA
Breastcancer	0.8833	0.6833	0.6667	0.6611
Exactly	0.8885	0.8885	0.9769	0.6692
Exactly_2	0.7500	0.6731	0.5615	0.6538
Breast_EW	0.8067	0.6483	0.6583	0.6767
Congress_EW	0.7313	0.5438	0.5875	0.5969
Heart_EW	0.8000	0.7115	0.7423	0.7192
Ionosphere_EW	0.7368	0.6353	0.5647	0.6618
Krvskp_EW	0.8487	0.8181	0.7792	0.7181
Lymphography	0.8472	0.7444	0.6222	0.6694
M-of-n	0.8808	0.9385	0.9115	0.6731
Sonar_EW	0.8425	0.7950	0.5667	0.7225
Spect_EW	0.7454	0.5977	0.5545	0.6182
Tic-tac-toe	0.8833	0.8889	0.8889	0.7056
Waveform_EW	0.8963	0.9650	0.8863	0.7663
Wine_EW	0.8308	0.9231	0.7269	0.7077
Zoo	0.7969	0.7250	0.6438	0.6344

Figures 3, 4, 5, provides an outline of the empirical results that gained from the different statistical measurements, best, worst; mean fitness. Results showed that SSA-SA approach is still performing better than native SSA and other optimisers in the literature.

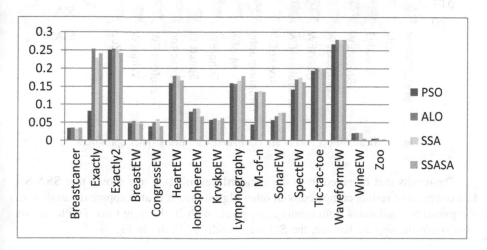

Fig. 3. Best fitness results are compared with many other optimizers

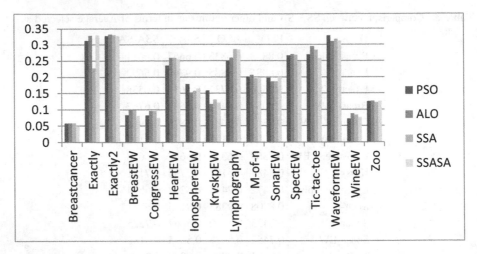

Fig. 4. Worst fitness results are compared with many other optimizers.

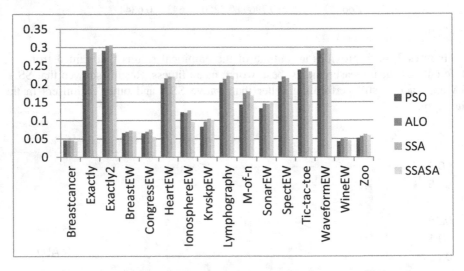

Fig. 5. Mean fitness results are compared with many other optimizers.

The results that obtained from the standard deviations (Std), shown that SSA-SA has competitive results compared with other algorithms. It is also apparent variation of the gained best solutions from running optimizers with 20 different runs. To illustration how stable the approaches are, the Std are introduced in the in Fig. 6.

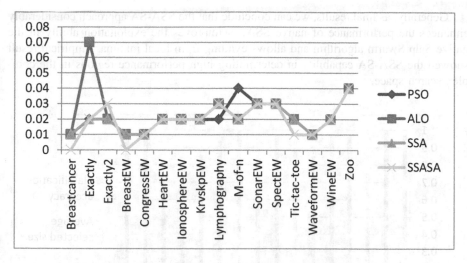

Fig. 6. Most remarkable solution in terms of standard deviation.

In this work, we used two large datasets, namely, Waveform. EW (5000 instance with 40 features) and Krvskp. EW (3196 instance with 36 features) which are high-dimensional datasets. In this term, the obtained Results from different measures indicated that the SSA-SA model clearly is outperforms other FS models in both datasets, illustrated in the following Fig. 7.

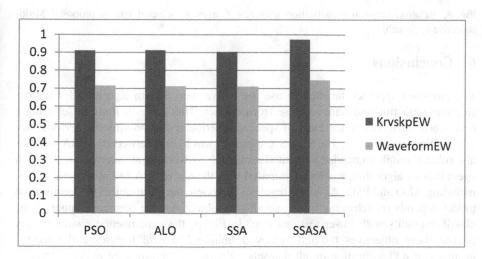

Fig. 7. Average performance over high-dimensional datasets.

Generally, as final results, we can conclude that the SSA-SA approach considerably enhances the performance of native SSA, is improves the exploitation ability in the native Salp Swarm algorithm and allows evading from local minima. Empirical result showed the SSA-SA capability in determining high performance regions in the complex search space.

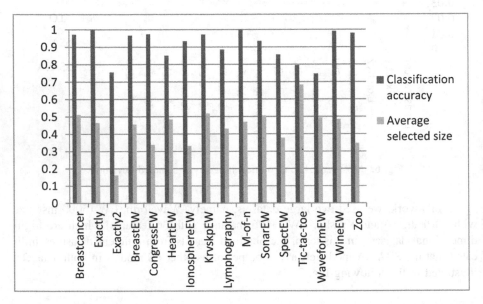

Fig. 8. Relation between classification accuracy & average selected size in proposed Multi-objective approach.

6   Conclusions

The proposed approach targeted to use the native Salp Swarm algorithm for dimensionality reduction and remove noise from dataset. This achieves through selecting a best subset attributes on the basis of specific criterion and improving the performance of classification accuracy. The SSA-SA approach was hybrid between the SSA with SA algorithm through embedded simulated annealing to work as an internal local search agent in SSA algorithm, and then compared with the native SSA and other approaches, including ALO and PSO. As an aggregation approach, the multi-objective optimisation model depends on achieving two contradictory objectives, are obtaining upper of the classifier quality with lower features size. In Fig. 8, the experimental results clearly proved these objectives through achieved, minimal size of features and maximal accuracy of a classification on all datasets. Obviously, the proposed hybrid approach can quickly expand the search spot by SA ability that utilizes to accept a worse quality solution than the current solution. The proposed approach can apply in some industry applications such as, for the fault diagnosis in wind turbine test rig datasets. Native Salp Swarm algorithm can be applied as a filter-based mode or aggregating with other local search algorithms at future works.

References

1. Liu, H., Motoda, H.: Feature Selection for Knowledge Discovery and Data Mining, vol. 454. Springer, Berlin (2012). https://doi.org/10.1007/978-1-4615-5689-3
2. Linoff, G.S., Berry, M.J.: Data Mining Techniques: For Marketing, Sales, and Customer Relationship Management. Wiley, New York (2011)
3. Shaker, K., Abdullah, S., Alqudsi, A.: Bacteria swarm optimisation approach for enrolment-based course timetabling problems. In: 7th Multidisciplinary International Conference on Scheduling. Theory and Applications, (MISTA 2015), pp. 515–525 (2015)
4. Abdullah, S., Shaker, K., Shaker, H.: Investigating a round robin strategy over multi algorithms in optimizing the quality of university course timetables. Int. J. Phys. Sci. **6**(6), 1452–1462 (2011)
5. Zhao, Z.A., Liu, H.: Spectral Feature Selection for Data Mining. CRC Press, Boca Raton (2011)
6. Liu, H., Motoda, H.: Feature Selection for Knowledge Discovery and Data Mining, vol. 454. Springer, Berlin (2012). https://doi.org/10.1007/978-1-4615-5689-3
7. Hadi, A.S. Fergus, P., Dobbins, C., Al-Bakry, A.M.: A machine learning algorithm for searching vectorised RDF data. In: 27th International Conference on Advanced Information Networking and Applications Workshops (WAINA), 25 March 2013, pp. 613–618. IEEE (2013)
8. Guyon, I., Gunn, S., Nikravesh, M., Zadeh, L.A.: Feature Extraction: Foundations and Applications, vol. 207. Springer, Berlin (2008). https://doi.org/10.1007/978-3-540-35488-8
9. Du, L., Swamy, S.: Search and Optimization by Metaheuristics, p. 434. Springer, New York City (2016). https://doi.org/10.1007/978-3-319-41192-7
10. Dhaenens, C., Jourdan, L.: Optimization and Big Data. Metaheuristics for Big Data, pp 1–21 (2016)
11. Diao, R., Shen, Q.: Nature inspired feature selection meta-heuristics. Artif. Intell. Rev. **44**(3), 311–340 (2015)
12. Bala, J., Huang, J., Vafaie, H., DeJong, K., Wechsler, H.: Hybrid learning using genetic algorithms and decision trees for pattern classification. IJCAI **1**, 719–724 (1995)
13. Meiri, R., Zahavi, J.: Using simulated annealing to optimize the feature selection problem in marketing applications. Eur. J. Oper. Res. **171**(3), 842–858 (2006)
14. Caruana, R., Freitag, D.: Greedy attribute selection. In: Machine Learning Proceedings, pp. 28–36 (1994)
15. Pacheco, J., Casado, S., Núñez, L.: A variable selection method based on Tabu search for logistic regression models. Eur. J. Oper. Res. **199**(2), 506–511 (2009)
16. Diao, R., Shen, Q.: Feature selection with harmony search. IEEE Trans. Syst. Man Cybern. B (Cybern.) **42**(6), 1509–1523 (2012)
17. Muni, P., Pal, R., Das, J.: Genetic programming for simultaneous feature selection and classifier design. IEEE Trans. Syst. Man Cybern. B (Cybern.) **36**(1), 106–117 (2006)
18. Alba, E., Garcia-Nieto, J., Jourdan, L., Talbi, G.: Gene selection in cancer classification using PSO/SVM and GA/SVM hybrid algorithms. In: IEEE Congress on Evolutionary Computation, CEC 2007. IEEE, pp. 284–290 (2007)
19. Nakamura, Y., Pereira, A., Costa, A., Rodrigues, D., Papa, P., Yang, S.: BBA: a binary bat algorithm for feature selection. In: Graphics, Patterns and Images (SIBGRAPI), 2012 25th SIBGRAPI Conference, IEEE, pp. 291–297 (2012)
20. Kanan, R., Faez, K.: An improved feature selection method based on ant colony optimization (ACO) evaluated on face recognition system. Appl. Math. Comput. **205**(2), 716–725 (2008)

21. Atyabi, A., Luerssen, M., Fitzgibbon, S., Powers, M.: Evolutionary feature selection and electrode reduction for EEG classification. In: Evolutionary Computation (CEC), IEEE Congress, IEEE, pp. 1–8 (2012)
22. Nemati, S., Basiri, M.E., Ghasem-Aghaee, N., Aghdam, H.: A novel ACO–GA hybrid algorithm for feature selection in protein function prediction. Expert Syst. Appl. 36(10), 12086–12094 (2009)
23. Oh, S., Lee, S., Moon, R.: Hybrid genetic algorithms for feature selection. IEEE Trans. Pattern Anal. Mach. Intell. 26(11), 1424–1437 (2004)
24. Talbi, G., Jourdan, L., Garcia-Nieto, J., Alba, E.: Comparison of population based metaheuristics for feature selection: application to microarray data classification. In: Computer Systems and Applications, 2008. AICCSA 2008. IEEE/ACS International Conference, pp. 45–52 (2008)
25. Khamees, M., Albakr, A., Shaker, K.: A new approach for features selection based on binary slap swarm algorithm. J. Theor. Appl. Inf. Technol. 96(7), 1896–1899 (2018)
26. Madin, P.: Field observations on the feeding behavior of salps (Tunicata: Thaliacea). Mar. Biol. 25(2), 143–147 (1974)
27. Mirjalili, S., Gandomi, H., Mirjalili, Z., Saremi, S., Faris, H., Mirjalili, M.: Salp swarm algorithm: a bio-inspired optimizer for engineering design problems. Adv. Eng. Softw. 114, 163–191 (2017)
28. Kirkpatrick, S., Gelatt, D., Vecchi, P.: Optimization by simulated annealing. Science 220 (4598), 671–680 (1983)
29. Lin, L., Fei, C.: The simulated annealing algorithm implemented by the MATLAB. Int. J. Comput. Sci. Issues (IJCSI) 9(6), 357 (2012)
30. Jensen, R., Shen, Q.: Finding rough set reducts with ant colony optimization. In: Proceedings of the 2003 UK Workshop on Computational Intelligence, vol. 1, No. 2, pp. 15–22 (2003)
31. Altman, S.: An introduction to kernel and nearest-neighbor nonparametric regression. Am. Stat. 46(3), 175–185 (1992)
32. Zawbaa, M., Emary, E., Parv, B.: Feature selection based on antlion optimization algorithm. In: Third World Conference on Complex Systems (WCCS), pp. 1–7. IEEE (2015)
33. Emary, E., Zawbaa, M., Hassanien, E.: Binary grey wolf optimization approaches for feature selection. Neurocomputing 172, 371–381 (2016)
34. Mafarja, M., Mirjalili, S.: Hybrid Whale Optimization Algorithm with simulated annealing for feature selection. Neurocomputing 260, 302–312 (2017)
35. Frank, A.: UCI Machine Learning Repository. University of California, Irvine. School of Information and Computer Science (2010). http://archive.ics.uci.edu/ml

Improving Multiobjective Particle Swarm Optimization Method

Intisar K. Saleh[1], Ufuk Özkaya[2], and Qais F. Hasan[1(✉)]

[1] Kirkuk Technical College, Northern Technical University, Mosul, Iraq
intisarks@gmail.com, dr.qaishasan@gmail.com
[2] Electronic and Communication Engineering Department,
Süleyman Demirel University, Isparta, Turkey
ufukozkaya@sdu.edu.tr

Abstract. An improved multiobjective particle swarm optimization algorithm is developed to get and compare Pareto fronts for constrained and unconstrained optimization test problems, with two objective functions and with a variable number of decision variables, available in literature. A new Minimum Angular Distance Information technique, to assign the best local guide for each particle within the swarm to get the Pareto front in the polar coordinate system, is adopted and verified. An external repository (archive) is used to store the nondominated particles at the end of each iteration, and a crowding distance technique is followed to maintain the archive size and the front diversity for each test problem. A self-adaptive penalty function technique is used to handle the constraint functions through transforming the original objective functions into new penalized functions based on their amount of constraint violation at each iteration. The developed algorithm is coded by Matlab formulas and verified via thirteen well-known test problems. Test results, represented in the regular Pareto fronts and the values of three comparative metrics (GD, S, and ER) calculated to verify the proposed algorithm, show more efficient and realistic agreements compared with that gained from previous studies and algorithms. Applying different engineering design problems to the developed algorithm is suggested as a future work.

Keywords: Multiobjective · PSO · Pareto front · Minimum angular distance

1 Introduction

In realistic design scope, an optimization problem may have more than one objective function (multiobjective) and may be with or without constraints. These objectives conflict between each other, which gives a large number of solutions called "Pareto front" that satisfy the requirements. In fact, each one of these solutions gives a priority of one objective over the others. In the last years, the evolutionary algorithms and strategies, for solving multiobjective optimization problems, have grown significantly [1], giving need to a wide variety of algorithms to solve many engineering design problems. These evolutionary algorithms present trends to develop and extend their applications. One of these trends is to enhance the efficiency of the algorithms and data structures used to keep nondominated vectors [2]. To maintain diversity of Pareto

© Springer Nature Switzerland AG 2018
S. O. Al-mamory et al. (Eds.): NTICT 2018, CCIS 938, pp. 143–156, 2018.
https://doi.org/10.1007/978-3-030-01653-1_9

fronts, researchers of multiobjective evolutionary optimization have produced some smart techniques (like Pareto Archive Evolutionary Strategy (PAES) adaptive grid [3]). Some other techniques are used to reduce size of population (like micro Genetic Algorithm (Micro-GA) [4]). Other techniques are used to handle unconstrained multiobjective optimization problems (like penalty functions [5]). Particle Swarm Optimization (PSO) is an evolutionary algorithm developed by Kennedy and Eberhart [6], and is a heuristic-based optimization technique. It is used to consider the social behaviour of flocking of birds and fish schooling. It is an efficient and simple population technique [7], therefore, it is extended by Coello Coello et al. [2] to deal with multiobjective optimization problems. Multiobjective PSO is modified, then, in two ways depending on dealing with the given objective functions. First, each objective function is treated separately, and second, all objective functions are evaluated for each particle [7].

PSO is an iterative technique, where each particle represents a potential solution and moves through the problem space following the optimum particle. To remember previous experience, each particle has a separate area of memory to store the best solution's position visited so far in the search space. This value is called as *pbest*. Another best value that is tracked by the particle swarm optimizer is the best value obtained by any particle in the neighbours of the particles. This value is called *lbest*. When a particle takes all the population as its topological neighbours, the best value is a global best and is called *gbest*. The particle swarm optimization concept consists of, at each time step, changing the velocity of each particle toward its *pbest* and *gbest*. After finding the best values, each particle updates its velocity and position accordingly [6]. Özkaya and Güneş [8] proposed a new method called "Minimum Angular Distance Information technique" to find *lbest* for each particle in the swarm to modify and improve multiobjective PSO. They suggested to implement this method to common benchmark functions in order to verify it, in addition, they recommended to overcome the diversity problem which may be raised when drawing the converged Pareto fronts. Therefore, the present study is aimed to develop and verify an improved, efficient, and realistic multiobjective PSO algorithm using new Minimum Angular Distance Information technique to solve different mathematical and engineering problems through establishing a regular set of solutions "Pareto front". In order to apply the algorithm for a wide range of various applications, it should have the ability to solve both constrained and unconstrained optimization problems having any number of decision variables through an efficient constraints handling technique. This algorithm, then, to be coded using Matlab functions and formulas and to be verified and compared using thirteen well-known test problems available in literature.

2   Multiobjective Particle Swarm Optimization Algorithm

An optimization problem could be an unconstrained straightforward problem. However, in more realistic problems, some constraints are imposed on the decision variables, this type is called "Constrained Optimization Problem" [9]. The most common approach in this field is the use of what is called "Penalty Function" which was originally proposed in the 1940s by Courant [10] and later expanded by Fiacco and

McCormick [11]. Self-adaptive penalty function is used where new dynamic penalized objective functions for the infeasible individuals can be calculated according to their constraints violation. A multiobjective optimization problem could be written in the form: minimize the objective vector $F(x) = [f_1(x), f_2(x), ..., f_k(x)]$ for k objective functions and for $x = [x_1, x_2, ..., x_n]^T$, the vector of decision variables. Our task is to determine the set of all objective vectors which satisfy all the constraints, the particular set of values $[x_1^*, x_2^*, ..., x_n^*]$ also yields multiple values for all the objective functions [12]. Therefore, instead of a unique optimum, there is a set of alternative trade-off solutions generally known as Pareto front, which represents the best compromise between the given objectives. A general multiobjective optimization includes a set of n parameters (decision variables), a set of k objective functions, and a set of m constraints.

In order to well define a multiobjective optimization algorithm, the search for the nondominated set of particles needs to define the concept of dominance and its related terms and techniques. An algorithm developed by Mishra and Harit [13] is considered in this study to store only nondominated solutions through a few comparisons to classify a particle into either dominated or nondominated. In single objective PSO, particles continue searching to get the optimum solution based on *gbest* and *pbest* until the expectation is met. The global best particle's guidance is essential for convergence to the minima/maxima. The presence of a set of nondominated solutions in multiobjective optimization redefines the guide selection, therefore, a new term called "best local guide" (*lbest*) is defined for each particle to determine Pareto solutions of the conflicting objectives. At the present study, Minimum Angular Distance Information technique is used to find the best local guide for each particle in the swarm [8], in which each dominated particle is guided by a nondominated particle having angular distance value close to its angular distance, as shown in Fig. 1. This is done by calculating the angular distance for all the particles at the current time-step and calculating to find, for each dominated particle, which nondominated particle having minimum angular difference in its polar coordinates. So, at each iteration and for each dominated particle there will be a nondominated particle assumed as its local guide leads it to Pareto front.

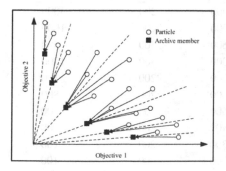

Fig. 1. Minimum Angular Distance Information technique for assigning *lbest* [8].

An external repository (archive) is used to store the nondominated particles added at each iteration. In order to maintain the capacity of this archive and to maintain the diversity "well distribution" of particles along the established Pareto front, a "Crowding Distance" technique is used, which is described by Raquel and Naval [14]. The crowding distance value of a solution provides an estimate of the density of solutions surrounding that solution [15]. In addition, to allow a quantitative assessment of our multiobjective optimization algorithm performance, three issues are taken into consideration [2]. These issues are: (Generational Distance (GD) [16], Spacing (S) [17], and Error Ratio (ER) [18]). Depending on the mentioned requirements, a Matlab code is developed and verified using thirteen well-known test problems available in literature with a stopping criteria of either finalizing total iterations or the total number of dominated particles at the end of any iteration becomes one. Also, a comparative study is done to examine efficiency of the proposed algorithm compared to other common algorithms.

3   Results and Discussion

The thirteen constrained and unconstrained multiobjective problems with two objective functions, considered to test the performance of the proposed algorithm, are summarized in Table 1 with some of the important input and output data for each one. In order to maintain the diversity in test problem selection, objective functions of different number of decision variables are selected with one, two, three, and ten variables. Four of these test problems are used by Coello Coello et al. [2], which found four different sets of results for each one of these problems, therefore, results of the present study are compared with these mentioned sets of results.

Table 1. Some important data for the thirteen investigated test problems.

Test problem	Size of whole swarm (particles)	Size of external repository (particles)	Total number of iterations (per one run)	Mean run time required (seconds)
Binh	1000	200	3000	122.4
Kursawe	1000	200	1000	148.1
Deb1	600	300	500	23.8
Deb2	1000	300	1200	157.3
Chackong & Haimes	300	200	100	11.65
Fonseca & Flemin.	600	200	800	55.87
Poloni	600	200	800	108.49
Schaffer N2	400	200	300	43.60
ZDT1	300	200	100	14.42
ZDT2	300	200	100	12.75
ZDT3	300	200	100	4.27
ZDT4	300	200	100	6.53
ZDT6	300	200	100	6.44

For the first example, a constrained multiobjective optimization test problem is used, which is proposed by Binh [19] and tested by Coello Coello et al. [2]. This test problem consists of minimizing two objective functions (f_1 and f_2) with three constraint functions (g_1, g_2, and g_3) and with two decision variables (x_i: where $1 \leq i \leq 2$). Graphical result for the Pareto front produced by the present algorithm show good agreement compared to the true Pareto front produced by Binh [19], as shown in Fig. 2 which shows also a comparison of these fronts with the algorithm produced by Coello Coello et al. (MOPSO) [2] and the other three well-known algorithms (NSGA-II, Micro-GA, and PAES). The results of the three comparative metrics (GD, S, and ER) are given through Tables 2 to 4. In the present study, the calculated metrics for this test problems and all other test problems represent the summary of running the Matlab code ten times per each test problem. These results are compared with the four sets of results given by Coello Coello et al. [2]. Denoting the minimum values with bold numbers, it is noticed that the best value for GD calculated by the present algorithm is slightly greater than that produced by the MOPSO algorithm, as given in Table 2, while the worst GD is lesser (better) than that produced by Coello Coello et al. [2] for all the sets of results. The average value of GD produced by the present algorithm is 0.1105 which is bigger than 0.036535 produced by the MOPSO algorithm, and the standard deviation of present calculated GD is 0.0113, which is better than 0.104459 produced by the MOPSO algorithm and better than that produced by the other mentioned algorithms.

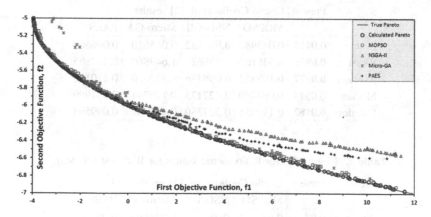

Fig. 2. Different Pareto fronts for Binh test problem.

Table 2. Result of Generational Distance metric values for Binh test problem.

GD	Present	Coello Coello et al. [2] results			
		MOPSO	NSGA-II	Micro-GA	PAES
Best	0.00966	**0.002425**	0.003885	0.00513	0.011321
Worst	**0.1139**	0.476815	0.678449	0.912065	0.919167
Average	0.1105	**0.036535**	0.084239	0.150763	0.193173
Median	0.1116	**0.007853**	0.011187	0.089753	0.033289
Std. dev.	**0.0113**	0.104459	0.165244	0.216558	0.249653

The results for S values, which are given in Table 3, show that the present calculated results give better values than that produced by the MOPSO algorithm in best, worst, average, median, and standard deviation terms. The average present calculated S is 0.0577, which is about half of 0.109452 produced by the MOPSO algorithm, and the standard deviation of the present calculated S is 0.0182 which is about 10% of the corresponding value given off by the MOPSO algorithm, which is 0.110051. The present calculated results of ER in all terms, which are given in Table 4, also show better values when compared with the four sets terms given by Coello Coello et al. [2]. The present average ER is 0.1145, which is less than 0.132532 given by MOPSO. In this problem, the total number of fitness particles used in the present algorithm to draw the Pareto front are 200 particles converged from 1000 particles used as the whole swarm, as shown in Table 1, while the number of converged fitness particles used by Coello Coello et al. [2] for the same problem was 5000, and this means that the present algorithm needs less fitness particles to get more precise and realistic Pareto front, and therefore, less time consuming. The mean time required for each full run of the thirteen test problems used in this study is shown in Table 1. For the current test problem, the mean time required to run the Matlab code with the given input data on a high performance microprocessor laptop is 122.4 s.

Table 3. Result of Spacing metric values for Binh test problem.

S	Present	Coello Coello et al. [2] results			
		MOPSO	NSGA-II	Micro-GA	PAES
Best	0.0414	0.043982	**0.001032**	0.065610	0.006669
Worst	**0.0938**	0.538102	1.48868	1.643860	0.432865
Average	**0.0577**	0.109452	0.098486	0.315020	0.110103
Median	0.0544	0.067480	**0.027173**	0.129744	0.081999
Std. dev.	**0.0182**	0.110051	0.327380	0.421742	0.099598

Table 4. Result of Error Ratio metric values for Binh test problem.

ER	Present	Coello Coello et al. [2] results			
		MOPSO	NSGA-II	Micro-GA	PAES
Best	**0.08**	**0.08**	0.75	0.734694	0.93
Worst	**0.14**	0.27	0.99	1.01639	1.01
Average	**0.1145**	0.132532	0.8965	0.927706	0.993
Median	0.155	**0.14**	0.92	0.936365	1.01
Std. dev.	**0.0188**	0.045007	0.067143	0.068739	0.025361

For the second example, an unconstrained multiobjective optimization test problem is used, which is proposed by Kursawe [20] and tested by Coello Coello et al. [2]. This test problem consists of minimizing two objective functions (f_1 and f_2) with three decision variables (x_i: where $1 \leq i \leq 3$). Graphical results for the present Pareto front show good agreement with the true Pareto front given by Kursawe [20], and compared to the fronts given by Coello Coello et al. [2], as shown in Fig. 3. It is noticed that all the comparison values (best, worst, average, and standard deviation) for the present calculated GD, as given in Table 5, are lesser than that given by Coello Coello et al. [2]. The standard deviation for the present GD is $2.91 * 10^{-4}$, which is about half of that given by the MOPSO algorithm. The present results for S, which are given in Table 6, show also lesser values than that produced by the MOPSO algorithm in best, worst, average, median, and standard deviation terms. Present standard deviation of S is 0.0095 which is much smaller than 0.01675 of the MOPSO algorithm. Present results of S, in most of their terms, show better performance than the results of other mentioned algorithms. Present results of ER, which are given in Table 7, also show better values when compared with the four sets of results given by Coello Coello et al. [2]. In this problem, the total number of fitness particles used to draw the present Pareto front are 200 particles converged from 1000 particles used as the whole swarm, as shown in Table 1, while the number of converged fitness particles used by Coello Coello et al. [2] for the same problem was 12000. Also, the mean time required to run the Matlab code with the given input data is 148.1 s.

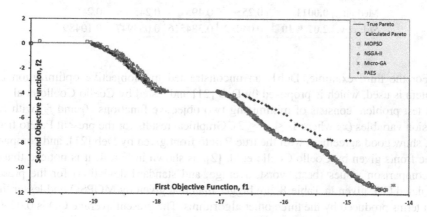

Fig. 3. Different Pareto fronts for Kursawe test problem.

Table 5. Result of Generational Distance metric values for Kursawe test problem.

GD	Present	Coello Coello et al. [2] results			
		MOPSO	NSGA-II	Micro-GA	PAES
Best	**0.0021**	0.00745	0.006905	0.006803	0.014670
Worst	**0.0032**	0.00960	0.103095	0.010344	0.157191
Average	**0.0024**	0.00845	0.029255	0.008456	0.549140
Median	**0.0024**	0.00845	0.017357	0.008489	0.049358
Std. dev.	**$2.91 * 10^{-4}$**	0.00051	0.027170	0.030744	0.030744

Table 6. Result of Spacing metric values for Kursawe test problem.

S	Present	Coello Coello et al. [2] results			
		MOPSO	NSGA-II	Micro-GA	PAES
Best	0.0368	0.06187	**0.018418**	0.071686	0.064114
Worst	**0.0562**	0.118445	0.065712	0.203127	0.340955
Average	0.0464	0.09747	**0.036136**	0.128895	0.197532
Median	0.0445	0.10396	**0.036085**	0.126655	0.186632
Std. dev.	**0.0059**	0.01675	0.010977	0.029932	0.064114

Table 7. Result of Error Ratio metric values for Kursawe test problem.

ER	Present	Coello Coello et al. [2] results			
		MOPSO	NSGA-II	Micro-GA	PAES
Best	**0**	0.18	0.06	0.18	0.10
Worst	**0.0015**	0.37	1.01	0.36	0.68
Average	**0.0012**	0.2535	0.56	0.27	0.27
Median	**0.0011**	0.255	0.495	0.245	0.245
Std. dev.	$\mathbf{2.07 * 10^{-4}}$	0.04082	0.384516	0.053947	0.10489

For the third example, Deb1, an unconstrained multiobjective optimization test problem is used, which is proposed by Deb [21] and tested by Coello Coello et al. [2]. This test problem consists of minimizing two objective functions (f_1 and f_2) with two decision variables (x_i: where $1 \leq i \leq 2$). Graphical results for the present Pareto front, also, show good agreement with the true Pareto front given by Deb [21], and compared to the fronts given by Coello Coello et al. [2], as shown in Fig. 4. It is noticed that all the comparison values (best, worst, average, and standard deviation) for the present calculated GD, given in Table 8, are bigger than that given by MOPSO and lesser than most terms produced by the three other algorithms. The present average GD is 0.02807, which is bigger than 0.000118 given by the MOPSO algorithm, and the standard deviation for the present GD is 0.03713, which is bigger than $8.61 * 10^{-5}$ given by the MOPSO algorithm and lesser than that given by the three other algorithms. The present calculated S terms, given in Table 9, show that they give bigger values than that produced by the MOPSO algorithm in best, worst, average, median, and standard deviation terms. The present average S is 0.03571, which is bigger than 0.010392 given by the MOPSO algorithm but lesser than that given by two other algorithms. The present standard deviation of S is 0.04868 which is bigger than 0.04868 by the MOPSO algorithm. The present results of ER, given in Table 10, show lesser values when compared with that given by MOPSO and the three other algorithms. The present average ER is 0.0672, which is lesser than 0.3335 given by the MOPSO algorithm and lesser than that given by the other algorithms. The present standard deviation of ER is 0.0118, which is lesser than that given by the other algorithms.

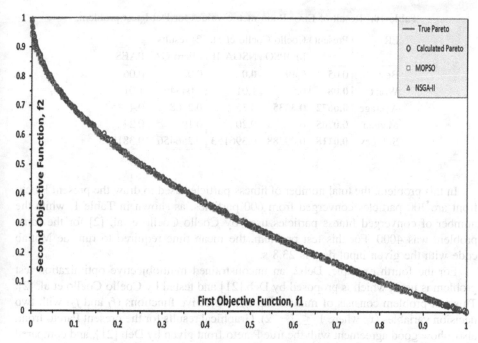

Fig. 4. Different Pareto fronts for Deb1 test problem.

Table 8. Result of Generational Distance metric values for Deb1 test problem.

GD	Present	Coello Coello et al. [2] results			
		MOPSO	NSGA-II	Micro-GA	PAES
Best	0.0058	**$8.61 * 10^{-5}$**	0.000133	$8.74 * 10^{-5}$	0.000114
Worst	0.053	**0.000191**	0.163146	0.811403	1.99851
Average	0.02807	**0.000118**	0.023046	0.047049	0.163484
Median	0.01281	**0.000111**	0.000418	0.000236	0.058896
Std. dev.	0.03713	**$8.61 * 10^{-5}$**	0.045429	0.18115	0.441303

Table 9. Result of Spacing metric values for Deb1 test problem.

S	Present	Coello Coello et al. [2] results			
		MOPSO	NSGA-II	Micro-GA	PAES
Best	0.0025	0.00727	**0.000205**	0.007596	0.009164
Worst	1.158	0.018676	**0.010234**	5.56727	19.8864
Average	0.03571	0.010392	**0.00369**	0.341659	1.114617
Median	0.094	0.009542	**0.002094**	0.2995	0.018755
Std. dev.	0.04868	0.002782	**0.003372**	1.247561	4.434594

Table 10. Result of Error Ratio metric values for Deb1 test problem.

ER	Present	Coello Coello et al. [2] results			
		MOPSO	NSGA-II	Micro-GA	PAES
Best	0.05	0.19	**0.0**	0.02	0.06
Worst	**0.08**	0.55	1.01	1.04545	1.01
Average	**0.0672**	0.3335	0.35	0.2568	0.4485
Median	**0.0705**	0.3	0.20	0.19	0.24
Std. dev.	**0.0118**	0.09388	0.396153	0.256456	0.381993

In this problem, the total number of fitness particles used to draw the present Pareto front are 300 particles converged from 600 particles, as shown in Table 1, while the number of converged fitness particles used by Coello Coello et al. [2] for the same problem was 4000. For this test problem, the mean time required to run the Matlab code with the given input data is 23.8 s.

For the fourth example, Deb2, an unconstrained multiobjective optimization test problem is used, which is proposed by Deb [21] and tested by Coello Coello et al. [2]. This test problem consists of minimizing two objective functions (f_1 and f_2) with two decision variables (x_i: where $1 \leq i \leq 2$). Graphical results for the present Pareto front, also, show good agreement with the true Pareto front given by Deb [21], and compared to the fronts given by Coello Coello et al. [2], as shown in Fig. 5. It is noticed that some of the comparison values (best, worst, and standard deviation) for the present GD, given in Table 11, are bigger than that given by MOPSO and lesser than some of the three other algorithm terms. The present average GD is 0.02155, which is better than that given by the MOPSO algorithm and the other algorithms, and the standard deviation for the present GD is 0.10231, which is bigger than that given by the MOPSO algorithm. The present results for S, given in Table 12, show bigger values than that produced by the MOPSO algorithm in best, worst, average, median, and standard deviation terms, but lesser values than most of the other algorithms. The present average S is 0.1692, which is bigger than 0.08358 given by the MOPSO algorithm, and the present standard deviation of S is 0.1405, which is bigger than 0.11821 of the MOPSO algorithm. The present results of ER, given in Table 13, show fluctuated values when compared with the values given by MOPSO and the three other algorithms. The present average ER is 0.2112, which is lesser than that given by the four algorithms. The present standard deviation for ER is 0.3899 is lesser than 0.400658 given by MOPSO and most of the other algorithms. In this problem, the total number of fitness particles used to draw the present Pareto front are 300 particles converged from 1000 particles used as the whole swarm, as shown in Table 1, while the number of converged fitness particles used by Coello Coello et al. [2] for the same problem was 10000. For this test problem, the mean time required to run the Matlab code with the given input data is 157.3 s.

All these notes mean that the proposed algorithm can be followed to represent the true Pareto front for the above four test problems more precisely and more accurately than other available well-known algorithms. Other good agreements are shown when testing an additional nine well-known test problems (Chakong and Haimes [22],

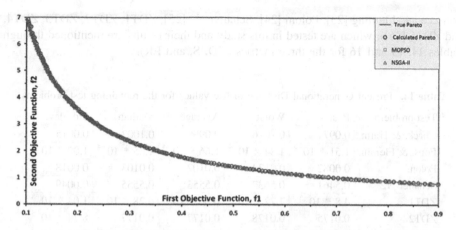

Fig. 5. Different Pareto fronts for Deb2 test problem.

Table 11. Result of Generational Distance metric values for Deb2 test problem.

GD	Present	Coello Coello et al. [2] results			
		MOPSO	NSGA-II	Micro-GA	PAES
Best	0.000684	**0.00043**	0.0007	0.000465	0.000453
Worst	0.2573	0.18531	0.208467	**0.183501**	0.221671
Average	**0.02155**	0.03273	0.044236	0.043466	0.194767
Median	0.0254	**0.00051**	0.000856	0.050042	0.070365
Std. dev.	0.10231	0.06062	0.07368	**0.048212**	0.204687

Table 12. Result of Spacing metric values for Deb2 test problem.

S	Present	Coello Coello et al. [2] results			
		MOPSO	NSGA-II	Micro-GA	PAES
Best	0.0967	0.04007	**0.026086**	0.030267	0.047844
Worst	0.8115	0.58185	**0.061422**	0.817642	0.664676
Average	0.1692	0.08358	**0.037447**	0.213584	0.194767
Median	0.0774	0.05494	**0.035529**	0.06301	0.070365
Std. dev.	0.1405	0.11821	**0.009238**	0.250586	0.204687

Table 13. Result of Error Ratio metric values for Deb2 test problem.

ER	Present	Coello Coello et al. [2] results			
		MOPSO	NSGA-II	Micro-GA	PAES
Best	0.09	**0.0**	0.02	0.08	0.02
Worst	**0.56**	1.01	1.01	1.01	1.01
Average	**0.2112**	0.25658	0.41450	0.25200	0.48900
Median	0.153	**0.045**	0.115	0.16	0.28
Std. dev.	0.3899	0.400658	0.459387	**0.231576**	0.438117

Fonseca and Fleming [23], Poloni [24], Schaffer N2 [25], ZDT1, ZDT2, ZDT3, ZDT4, and ZDT6 [26]), which are tested in this study and their results are mentioned through Tables 14, 15 and 16 for the three metrics (GD, S, and ER).

Table 14. Present Generational Distance metric values for the remaining test problems.

Test problem	Best	Worst	Average	Median	Std. dev.
Chack. & Haim.	0.0971	0.1016	0.0999	0.1003	0.0015
Fons. & Flemin.	$1.51 * 10^{-4}$	$1.84 * 10^{-4}$	$1.66 * 10^{-4}$	$1.65 * 10^{-4}$	$1.02 * 10^{-5}$
Poloni	0.0082	0.0134	0.0107	0.0103	0.0018
Schaffer N2	0.5461	0.5637	0.5553	0.5555	0.0049
ZDT1	$4.8 * 10^{-5}$	$7.72 * 10^{-5}$	$6.26 * 10^{-5}$	$6.28 * 10^{-5}$	$1.04 * 10^{-5}$
ZDT2	0.0175	0.0178	0.0177	0.0177	$8.74 * 10^{-5}$
ZDT3	$1.07 * 10^{-4}$	$1.34 * 10^{-4}$	$1.17 * 10^{-4}$	$1.17 * 10^{-4}$	$8.75 * 10^{-5}$
ZDT4	$3.93 * 10^{-5}$	$6.81 * 10^{-5}$	$5.62 * 10^{-5}$	$5.73 * 10^{-5}$	$9.57 * 10^{-6}$
ZDT6	$2.44 * 10^{-5}$	$2.63 * 10^{-5}$	$2.18 * 10^{-5}$	$2.56 * 10^{-5}$	$9.48 * 10^{-6}$

Table 15. Present Spacing metric values for the remaining test problems.

Test problem	Best	Worst	Average	Median	Std. dev.
Chack. & Haim.	0.4339	0.909	0.5807	0.5521	0.135
Fons. & Flemin.	0.0021	0.004	0.0034	0.0034	$3.58 * 10^{-4}$
Poloni	0.031	0.0413	0.037	0.0369	0.0036
Schaffer N2	0.0135	0.0189	0.0153	0.0166	0.0046
ZDT1	0.0018	0.00198	0.00185	0.0019	$5.25 * 10^{-5}$
ZDT2	0.0017	0.00191	0.0018	0.0018	$7.2 * 10^{-5}$
ZDT3	0.0018	0.00199	0.0019	0.0019	$7.1 * 10^{-5}$
ZDT4	0.0017	0.0019	0.0018	0.0019	$7.44 * 10^{-5}$
ZDT6	0.00137	0.0015	0.0014	0.0014	$5.48 * 10^{-5}$

Table 16. Present Error Ratio metric values for the remaining test problems.

Test problem	Best	Worst	Average	Median	Std. dev.
Chack. & Haim.	0.4339	0.909	0.5807	0.5521	0.135
Fons. & Flemin.	0.0021	0.004	0.0034	0.0034	$3.58 * 10^{-4}$
Poloni	0.031	0.0413	0.037	0.0369	0.0036
Schaffer N2	0.0135	0.0189	0.0153	0.0166	0.0046
ZDT1	0.0018	0.00198	0.00185	0.0019	$5.25 * 10^{-5}$
ZDT2	0.0017	0.00191	0.0018	0.0018	$7.2 * 10^{-5}$
ZDT3	0.0018	0.00199	0.0019	0.0019	$7.1 * 10^{-5}$
ZDT4	0.0017	0.0019	0.0018	0.0019	$7.44 * 10^{-5}$
ZDT6	0.00137	0.0015	0.0014	0.0014	$5.48 * 10^{-5}$

4 Conclusions and Recommendations

In this study, a multiobjective PSO algorithm with two objective functions is developed and improved for solving constrained and unconstrained test problems having a different number of decision variables. A new Minimum Angular Distance Information technique is used to assign the best local guide to each particle within the swarm to get the Pareto front in the polar coordinates system. A Matlab code, then, is written and executed following the steps put to achieve this algorithm. Thirteen well-known test problems are used to verify the proposed algorithm, most of their results are discussed and compared with other algorithms and studies, and the following conclusions are presented:

The new improved algorithm can well represent the Pareto fronts for multiobjective constrained and unconstrained optimization test problems, with a different number of decision variables. All the techniques and algorithms used to construct the improved algorithm (like Minimum Angular Distance Information, determination of nondominated set of particles, archive maintenance, and self-adaptive penalty function) are efficient to represent the correct and realistic Pareto front for all the test problems.

The present calculated results of the three comparative metrics (GD, S, and ER), show improved fluctuations when compared with the same corresponding results given in literature. The developed algorithm shows efficient response represented in less number of fitness particles used to draw a more accurate and realistic Pareto front compared with previous well-known algorithms.

A lot of researches can be suggested to extend the developed algorithm to be applicable for different engineering design problems, especially that are suffering from constraint requirements with multidimensional decision variables.

References

1. Coello Coello, A.: Theoretical and numerical constraint-handling techniques used with evolutionary algorithms: a survey of the state of the art. J. Comput. Methods Appl. Mech. Eng. **191**(11–12), 1245–1287 (2002)
2. Coello Coello, A., Pulido, T., Lechugu, M.: Handling multiple objectives with particle swarm optimization. IEEE Trans. Evol. Comput. **8**(3), 265–279 (2004)
3. Knowles, J., Corne, W.: Approximating the nondominated front using the Pareto archived evolution strategy. J. Evol. Comput. **8**(2), 149–172 (2000)
4. Coello Coello, A., Pulido, T.: Multiobjective optimization using a micro-genetic algorithm. In: Spector, L., et al. (eds.) Proceedings of Genetic and Evolutionary Computation Conference (GECCO 2001), San Francisco, CA (2001)
5. Tessema, G.: A self adaptive genetic algorithm for constrained optimization. M. Sc. thesis, Faculty of Graduate College, Oklahoma State University, pp. 29–39, December 2006
6. Kennedy, J., Eberhart, R.: Particle swarm optimization. In: Proceedings of the Fourth IEEE International Conference on Neural Networks, Perth, Australia (1995). IEEE Service Center 1942–1948, 1995
7. Kumer, V., Minz, S.: Multi-objective particle swarm optimization: an introduction. J. Smart Comput. Rev. **4**(5), 335–353 (2014)

8. Özkaya, U., Güneş, F.: A modified particle swarm optimization algorithm and its application to the multiobjective FET modeling problem. Turk J. Elec. Eng. Comput. Sci. **20**(2), 263–271 (2012)
9. Michalewicz, Z., Schoenauer, M.: Evolutionary algorithms for constrained parameter optimization problems. J. Evol. Comput. **4**(1), 1–32 (1996)
10. Courant, R.: Variational methods for the solution of problems of equilibrium and vibration. Bull. Am. Math. Soc. **49**(1), 1–23 (1943)
11. Fiacco, V., McCormick, P.: Nonlinear Programming: Sequential Unconstrained Minimization Techniques. Wiley, New York (1968). 201 p
12. Coello Coello, A.: comprehensive survey of evolutionary-based multiobjective optimization techniques. Int. J. Knowl. Inf. Syst. **1**(3), 269–308 (1999)
13. Mishra, K., Harit, S.: A fast algorithm for finding the nondominated set in multiobjective optimization. Int. J. Comput. Appl. (0975-8887) **1**(25), 35–39 (2010)
14. Raquel, C., Naval, P.: An effective use of crowding distance in multiobjective particle swarm optimization. In: GECCO 2005 Proceedings of the 7th Annual Conference on Genetic and Evolutionary Computation, 25–29 June 2005, Washington DC, USA, pp. 257–264 (2005)
15. Deb, K., Agrawal, S., Pratap, A., Meyarivan, T.: A fast Elitist non-dominated sorting genetic algorithm for multi-objective optimization: NSGA-II. In: Schoenauer, M., et al. (eds.) PPSN 2000. LNCS, vol. 1917, pp. 849–858. Springer, Heidelberg (2000). https://doi.org/10.1007/3-540-45356-3_83
16. Van Veldhuizen, D.A., Gary, B.: Multiobjective evolutionary algorithm research: a history and analysis. Technical report TR-98-03, Department of Electrical and Computer Engineering, Graduate School of Engineering, Air Force Institute of Technology, Wright-Patterson AFB, Ohio (1998)
17. Schott, J.: Fault tolerant design using single and multicriteria genetic algorithm optimization. Master's thesis, Department of Aeronautics and Astronautics, Massachusetts Institute of Technology, Cambridge, MA (1995)
18. Van Veldhuizen, D.A.: Multiobjective evolutionary algorithm: classification, analysis, and new innovations. Ph.D. thesis, Department of Electrical and Computer Engineering, Graduate School of Engineering, Air Force Institute of Technology, Wright-Patterson AFB, Ohio (1999)
19. Binh, T.: A multi objective evolutionary algorithm: the study cases. Technical report, Institute for Automation and Communication, Barleben, Germany (1999)
20. Kursawe, F.: A variant of evolution strategies for vector optimization. In: Schwefel, H.-P., Männer, R. (eds.) PPSN 1990. LNCS, vol. 496, pp. 193–197. Springer, Heidelberg (1991). https://doi.org/10.1007/BFb0029752
21. Deb, K.: Multi-objective genetic algorithms: problem difficulties and construction of test problems. J. Evol. Comput. **7**(1), 205–230 (1999)
22. Chakong, V., Haimes, Y.: Multiobjective Decision Making Theory and Methodology. Elsevier Science, New York (1983)
23. Fonseca, C., Fleming, P.: Genetic algorithms for multi-objective optimization: formulation, discussion and generalization. In: Proceedings of the 5th International Conference on Genetic Algorithms, San Francisco, CA, USA (1993)
24. Poloni, C., Giurgevich, A., Onesti, A., Pediroda, V.: Hybridization of a multi-objective genetic algorithm, a neural network and a classical optimizer for a complex design problem in fluid dynamics. J. Comput. Methods Appl. Mech. Eng. **186**(2–4), 403–420 (2000)
25. Schaffer, J.: Multiple objective optimization with vector evaluated genetic algorithms. In: Proceedings of the First International Conference on Genetic Algorithms, pp. 93–100 (1985)
26. Zitzler, E., Deb, K., Theile, L.: Comparison of multiobjective evolutionary algorithms: empirical results. J. Evol. Comput. **8**(2), 173–195 (2000)

Intelligent Control System

Implementation of Self Tune Single Neuron PID Controller for Depth of Anesthesia by FPGA

Layla Hattim[1](✉), Ekhlas H. Karam[2], and Abbas Hussain Issa[3]

[1] Control and System Engineering Department,
University of Technology, Baghdad, Iraq
layhatim@yahoo.com

[2] Computer Engineering Department, University of Al Mustansyria,
Baghdad, Iraq
ek_karam@yahoo.com

[3] Electrical Engineering Department, University of Technology, Baghdad, Iraq
30050@uotechnology.edu.iq

Abstract. In general intravenous anesthesia, drugs must be given in a proper rate to ensure the safety of the patient during the surgical operation. Pharmacokinetic-Pharmacodynamic (PK/PD) model was used to represent the relationship between the given drug and its effect on the patient. In this paper and depending on the PK/PD model, a simple control scheme is suggested to control the Depth of Anesthesia (DOA). This control scheme composed from the single neuron self-tune PID controller, in addition to use the exact linearization strategy to overcome the nonlinearity of the PK/PD model due to the Hill function. The input to closed loop controlled system is the propofol and one output is the Bispectral index (BIS) reading for the patient from monitoring device during the surgery.

The purpose of use the single neuron PID controller is to generate an optimal infusion rate as a control signal by self-adjustment for PID controller parameters and then implement it on FPGA Spartan Kit XC3SD3400ACS484-4 by using Xilinx development tool (ISE 14.6). The Simulation results show the efficiency of the suggested control scheme in tracking the desired BIS for all patients' cases.

Keywords: Single neuron · PID controller · PK/PD model · Bispectral index (BIS) · FPGA

1 Introduction

At present, anesthesia is an essential component of any surgery to ensure the safety of the patient during the surgical operation. Automatically controlling methods are very important to do this by calculate the suitable amount needed to make the patient in a

© Springer Nature Switzerland AG 2018
S. O. Al-mamory et al. (Eds.): NTICT 2018, CCIS 938, pp. 159–170, 2018.
https://doi.org/10.1007/978-3-030-01653-1_10

suitable level for anesthesia. A good way for measuring the status of the patient under anesthesia among all the different ways found in these years is the Bispectral index (BIS), it is one of the best monitoring index used [1].

In biomedical engineering studies, there are many things that add difficulties to the designing of any controller, like the human nature's that vary from one patient to other, noticeable changes in the reaction to any effects, the response of the patient that is always reflects nonlinearity and different variables can effect on the system, because of all these reasons, intelligent adaptive control systems are very necessary to dominate like these systems because these controllers can be deal with the uncertain complex systems without needing to the system mathematical model [2].

Different adaptive intelligent controller have been suggested to control the depth of anesthesia (DOA) such as adaptive fuzzy logic control (AFLC) [2, 3], adaptive neural networks (ANN) [4], adaptive neuro-fuzzy inference system (ANFIS) [5], nonlinear backstepping controller [6], extended dynamic matrix control (EDMC) [7], adaptive model predictive control [8], adaptive fuzzy sliding mode control (AFSMC) [9], all these approaches have shown their efficacy in controlling the depth of anesthesia.

In this paper, a simple control scheme is suggested to control the DOA, this scheme includes the use of the single neuron auto-tune PID which is designed by [10] as a feed forward controller and the exact linearization strategy that proposed by [11] to deal with the nonlinearity of the Hill equation. Also in this paper, the single neuron auto-tune PID which is used to determine the optimal control signal (the infusion rate) is designed by the Field Programmable Gate Arrays (FPGA) then implemented on real FPGA Spartan Kit.

The paper is organized as listed: The patient mathematical model is explained in Sect. 2. In Sect. 3, the proposed controller is demonstrated. The implementation and the simulation are given in Sect. 4. In Sect. 5, the conclusions are presented.

2 Patient Mathematical Model

The model that represent the human body after taking the drug and show its effect on the body and then monitor the status on the BIS monitoring device is the Pharmacokinetic-Pharmacodynamic (PK/PD) model that translate the Propofol drug effect on the human body parts, is based on two part; a PK model for adopting the delivery of drug to all body parts, and a PD model to describe the body response as a value on BIS monitoring device [12]. The integrated PK/PD structure consists of a three parts for describe PK model with an effect part, the central part (intravascular blood) name $V1$, a rapid peripheral (muscle) and a slow peripheral (fat) names are $V2$ and $V3$, are tied to the central part, the transfer of the drug between all parts is done by the constant rate ($k12$, $k21$, $k13$, $k31$) as shown in Fig. 1 [13, 14].

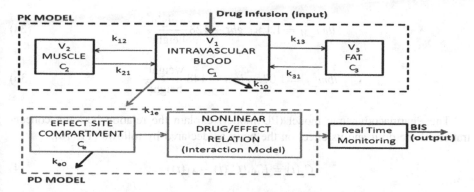

Fig. 1. Compartmental model of the patient.

The linear equations of PK model is expressed by [11–14]:

$$C_1'(t) = -[K_{10} + K_{12} + K_{13}].C_1(t) + K_{21}.C_2(t) + K_{13}.C_3(t) + u(t)/V$$

$$C_2'(t) = K_{12}.C_1(t) - K_{21}.C_2(t) \tag{1}$$

$$C_3'(t) = -K_{13}.C_1(t) + K_{31}.C_3(t)$$

where C1 is the drug concentration in the central compartment. The concentrations of drug in peripheral compartments are C2 and C3 respectively and in the effect-site compartment by Ce. The variable kij for i = 0:3, i ≠ j, is drug transfer from the ith to the jth compartment, k1e the drug transfer from central compartment to the effect site-compartment, k10 and k1e is the extraction drug variable from the central compartment and the effect-site compartment, u(t) is anesthesia drug rate needed. The parameters kij of PK models depend on patient age, weight, height and gender. For Propofol, the kij parameters can be and calculated by [6, 11, 15]:

$$V_1 = 4.27[l], V_2 = 18.9 - 0.391.(age - 53)[l], V_3 = 238[l]$$
$$C_{l1} = 1.89 + 0.456(weight - 77) - 0.0681(lbm - 59) + 0.264(height - 177)[l/min]$$

$$C_{l2} = 1.29 - 0.024(age - 53)[l/min], C_{l3} = 0.836[l/min] \tag{2}$$

$$K_{10} = C_{l1}/V_1 min^{-1}, K_{12} = C_{l2}/V_1 min^{-1}, K_{13} = C_{l3}/V_1 min^{-1}$$
$$K_{21} = C_{l2}/V_2 min^{-1}, K_{31} = C_{l3}/V_3 min^{-1}, K_{e0} = 0.456 min^{-1}$$

The Cl1 is the rate that extracted from the body, and Cl2 and Cl3 are the rates that extracted from central compartment to other compartments by distribution. The lean body mass (lbm) for men (M) and women (F) are calculated by [11]:

$$lbm\_M = 1.1.weight - 128\frac{weight^2}{height^2} \tag{3a}$$

$$lbm\_F = 1.07.weight - 148\frac{weight^2}{height^2} \tag{3b}$$

The Pharmacodynamic model (PD model) explain the relation between concentration of the drug and its effect on the body as declared by follows [11, 16]:

$$C'_e(t) = K_{e0}.(C_e(t) - x_1(t)) \tag{4}$$

$$BIS(t) = E_0 - E_{max}\frac{C_e(t)^\gamma}{C_e(t)^\gamma + EC_{50}^\gamma} \tag{5}$$

The BIS value depend on the scaling band specified according to the patient's status, this scaling band specify the band from (100–80) as a fully awake, while band of (60–40) explain the suitable value for completing the surgery safely and finally band of value from (40–20) translate deep anesthesia state which is mean very dangerous level [6]. BIS is associated with Ce by a nonlinear equation called Hill equation. E0 means the fully awake state and assigned to 100, Emax will reflects the fully effect of drug, EC50 is the half drug effect and indication of response of patient to drug, while γ is the curve steepness. The inverse of the Hill curve is explained by [11, 16]:

$$C_e(t) = EC_{50}.\left(\frac{E_0 - BIS(t)}{E_{max} - E0 + BIS(t)}\right)^{1/\gamma} \tag{6}$$

At this point the patient model (PK/PD) model is demonstrated clearly.

3   The Suggested Control Scheme

PID controllers are classified as a flexible and popular controller for improving the performance of many systems. Recently, the PID controller's design and tuning not like the original one and the design will digitally designed. These new digital controllers include many ways to enhance their performance [17], and one of these way is the use of neural network for tuning the PID controller parameters.

The block diagram for our suggested control scheme for the DOA is shown in Fig. 2, this control scheme including use the single neuron self-tune PID controller (which designed by [10]) as forward controller, in addition to the exact linearization strategy (that proposed by [11]) to overcome the nonlinearity of the Hill function (Eq. (5)) which complicated use the single neuron PID controller. The exact linearization strategy is based on Hill nonlinearity compensate by generate the inverse Hill equation (Eq. (6)) due to the Hill nonlinearity is a monotonic function [11].

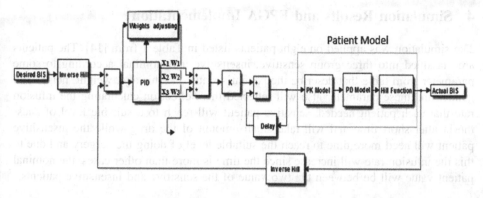

Fig. 2. The block diagram for the suggested control scheme for DOA.

The neural network structure translate the biology nerve system networks that it have input-output system. Characteristics of the neural network is achieved from artificial neurons or nodes connected to form a network, network type or structure and the way of the training and updating algorithm [18]. As shown from Fig. 2, the weights of the net [w1, w2, w3] will be the gains of the single neuron PID controller, and the proportional error x1(k), the integral error x2(k) and the derivative error x3(k) are given by [10]:

$$x_1(k) = e(k) - e(k-1)$$
$$x_2(k) = e(k)$$
$$x_3(k) = e(k) - 2e(k-1) + e(k-2)$$
(7)

The output of the single neuron which represent the control signal (infusion rate) is given by:

$$u(k) = u(k-1) + K \sum_{i=1}^{3} (w_i(k).x_i(k))$$
(8)

K is a gain to fast or slow the system response, the weights updating will done by supervised delta learning algorithm as shown by [10]:

$$w1(k) = w1(k-1) + \eta_p e(k-1)u(k-1)$$
(9a)

$$w2(k) = w2(k-1) + \eta_I e(k-1)u(k-1)$$
(9b)

$$w3(k) = w3(k-1) + \eta_D e(k-1)u(k-1)$$
(9c)

ΠP, ΠI, ΠD are the learning speeds for proportional, integrator, derivative for PID controller respectively [10, 19].

4   Simulation Results and FPGA Implementation

The simulation was applied on eight patients listed in Table 1 from [14]. The patients are classified into three group sensitive, insensitive and nominal according to some parameter from table that describe the patient's status like his age, weight, height and gender and these parameter also will reflected there effect on calculating the infusion rate that each patient needed, sensitive patient will reach to a suitable level of anesthesia after short time and will take small amount of the drug while the insensitive patient will need more time to reach the suitable level of doing the surgery and due to this the infusion rate will increase since the time is more than other cases, the nominal patient value will be between the two value of the sensitive and insensitive patients.

Table 1. Characteristic variables for patients [14].

Patient	Age	Height	Weight	Sex	EC50	E0	Emax	γ
1	40	163	54	F	6.33	98.8	94.1	2.24
2	36	163	50	F	6.76	98.6	86	4.29
3	28	164	52	F	8.44	91.2	80.7	4.1
4	50	163	83	F	6.44	95.9	102	2.18
5	28	164	60	M	4.93	94.7	85.3	2.46
6	43	163	59	F	12.1	90.2	147	2.42
7	37	187	75	M	8.02	92	104	2.1
8	38	174	80	F	6.56	95.5	76.4	4.12

As a result to apply the suggested control scheme to the plant (Patient Model) which is derived from patient variables it mean from its age, weight, height and gender, the response will be as indicated in Fig. 3; if we analyze this figure we see that according to anesthesia stages (induction, maintenance, recovery), the induction stage will translate the status of patient from awake state until starting to reach to the level that allow to make the surgery, it starts from 110 s it mean approximately 2 min this for sensitive patient while the insensitive patient will start in 175 s it mean after approximately 3 min, it seems to be not different but in fact it effected on the infusion rate since it will be increased each second as indicated in Fig. 4, while the maintenance stage will start when the value of BIS seems to be not change and ensue to stay at this point till the end of the surgery, so from Fig. 3 we see the settling time will be at 3 min after this the system still stable without any undershoot or overshot to reach to the steady state case and still in that value.

Finally the error will calculated according to the difference between desired value and actual value for all patients as indicated (see Fig. 5).

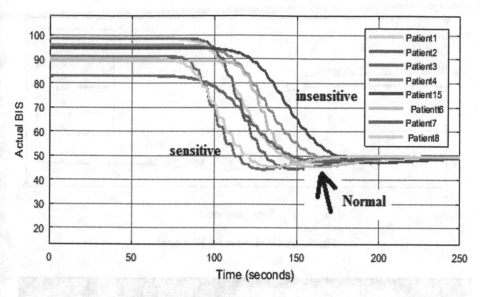

Fig. 3. Actual BIS reading for patients.

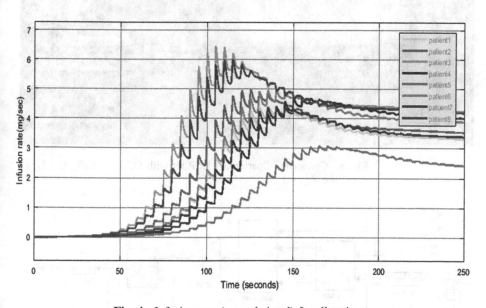

Fig. 4. Infusion rate (control signal) for all patients.

At this point after testing the patients response and calculated the depth of anesthesia drug, the used single neuron PID controller must be converted to VHDL code in order to be implemented on a Xilinx Spatran Kit XC3SD3400ACS484-4 (see Fig. 6.), this will done by using the Xilinx system generator from Xilinx development tool Integrated Software Environment (ISE) version 14.6 by connect the proposed controller (see Figs. 7 and 8).

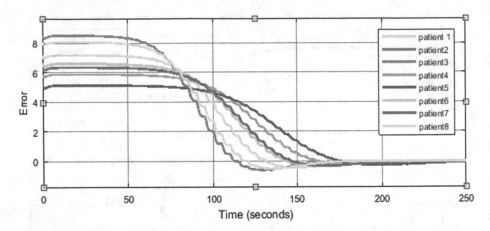

Fig. 5. Tracking error for all patients

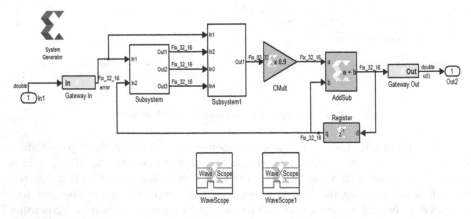

Fig. 6. Connecting Spartan-3A DSP kit with PC.

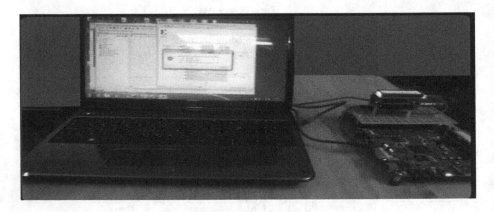

Fig. 7. Implementation of the single neuron PID controller by Xilinx system generator.

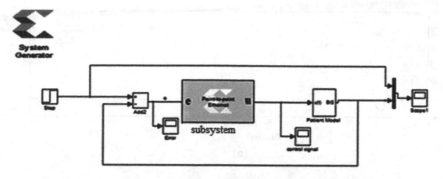

Fig. 8. FPGA connection to PC.

Table 2. Table captions should be placed above the tables.

Device utilization summary				
Logic utilization	Used	Available	Utilization (%)	Note(s)
Number of slice flip flops	352	47,744	1	
Number of 4 input LUTs	1,042	47,744	2	
Number of occupied slices	770	23,872	3	
Number of slices containing only related logic	770	770	100	
Number of slices containing unrelated logic	0	770	0	
Total number of 4 input LUTs	1,183	47,744	2	
Number used as logic	1,042			
Number used as a route-thru	141			
Number of bonded IOBs	65	309	21	
Number of BUFGMUXS	1	24	4	
Number DSP48As	42	126	33	
Average fanout of Non-Clock Nets	1.54			

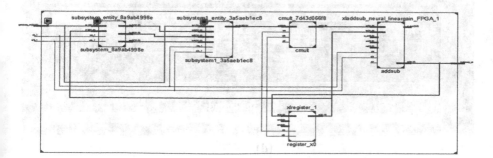

Fig. 9. RTL scheme of the controller

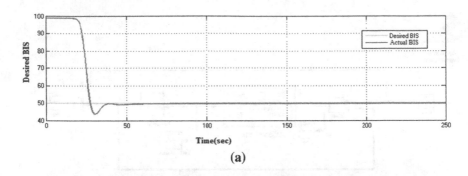

(a)

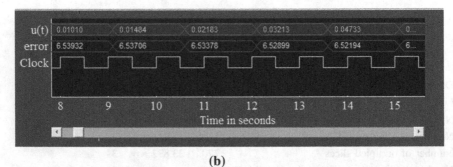

(b)

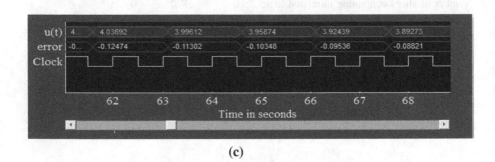

(c)

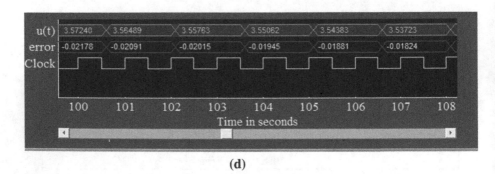

(d)

Fig. 10. (a) output system response; (b) Infusion drug when starting; (c) Infusion drug after taking the drug (induction stage); (d) Infusion drug when be in anesthesia process completely (maintenance stage).

From this circuit we can obtain VHDL code by run generation to see the netlist generated that contain the code file and use it to see the design summary and Know the size of the circuit it mean how much it utilized from the FPGA kit as in Table 2, also see the RTL scheme for the circuit as in Fig. 9.

Then after testing the circuit, the system response (BIS) and control signal ($u(t)$) for the one patient are as indicated (see Fig. 10).

5   Conclusion

This paper present FPGA implementation for the single neuron self-tune PID controller. This controller is consider as apart from simple suggested control scheme which also use the exact linearization strategy to overcomes the nonlinearity of the Hill function in PK/PD model which complicated use the single neuron PID controller. The mathematical model PK/PD model is based on patients characteristics and considering the BIS value is the verify of patient status under surgery, the suggested control scheme provide stable response in spite of the difference in patients parameters. FPGA kits are very efficient tool that provide more flexibility and easy to implement on it, with the use of system generator, VHDL code is generated and the system summary also can provided, hardware FPGA design for the single neuron PID controller is implemented on Spartan-3A DSP XA3SD3400A Xilinx kit, the maximum frequency allowed for this controller is 10.465 MHz.

References

1. Rezvaniana, S., Towhidkhaha, F., Ghahramania, N., Rezvanianb, A.: Increasing robustness of the anesthesia process from difference patient's delay using a state-space model predictive controller. Proc. Eng. **15**, 928–932 (2011)
2. Chou, Y.C., Abbod, M.F., Shieh, J.S., Hsu, Y.H.: Multivariable fuzzy logic/self-organizing for anesthesia control. J. Med. Biol. Eng. **30**(5), 297–306 (2010)
3. Araujo, H., Xiao, B., Liu, C., Zhao, Y., Lam, H.K.: Design of type-1 and interval type-2 fuzzy PID control for anesthesia using genetic algorithms. J. Intell. Learn. Syst. Appl. **6**, 70–93 (2014)
4. Haddad, W.M., Bailey, J.M., Hayakawa, T., Hovakimyan, N.: Neural network adaptive output feedback control for intensive care unit sedation and intraoperative anesthesia. IEEE Trans. Neural Netw. **18**(4), 1049–1066 (2007)
5. Shalbaf, A., Saffar, M., Sleigh, J.W., Shalbaf, R.: Monitoring the depth of anesthesia using a new adaptive neuro-fuzzy system. IEEE J. Biomed. Health Inf. **22**, 671–677 (2016)
6. Khaqan, A., Bilal, M., Ilyas, M., Ijaz, B., Riaz, R.A.: Control law design for propofol infusion to regulate depth of hypnosis: a nonlinear control strategy. Comput. Math. Methods Med. **2016**, 1–10 (2016). Hindawi Publishing Corporation Article ID 1810303
7. Bamdadian, A., Towhidkhah, F., Marami, B.: Controlling the depth of anesthesia by using extended DMC. In: Proceedings of the 2008 IEEE, CIBEC 2008 (2008)
8. Yelneedi, S., Lakshminarayanan, S., Rangaiah, G.P.: Adaptive model predictive control for automatic regulation of anesthesia by simultaneous administration of two intravenous drugs. In: Conference AIChE Annual Meeting, Group Computing and Systems Technology Division (2007)

9. Hosseini, S.H.S., Khazaei, M., Khomarlou, Z.A., Geramipour, A.H.: Controlling the depth of anesthesia using adaptive fuzzy sliding mode control strategy. IJMEC **5**(16), 2313–2326 (2015)

10. Liu, X: Single neuron self-tuning PID control for welding molten pool depth. In: Proceedings of the 7th World Congress on Intelligent Control and Automation 25–27 June, Chongqing, China (2008)

11. Naşcu, I.: Advanced multiparametric optimization and control studies for anaesthesia. A thesis submitted to Imperial College London for the degree of Doctor of Philosophy, Centre for Process Systems Engineering, Department of Chemical Engineering, Imperial College London, United Kingdom, January 2016

12. Yelneedi, S., Samavedham, L., Rangaiah, G.P.: Advanced control strategies for the regulation of hypnosis with propofol. Am. Chem. Soc. Ind. Eng. Chem. Res. **48**(8), 3880–3897 (2009)

13. Ilyas, M., et al.: A review of modern control strategies for clinical evaluation of propofol anesthesia administration employing hypnosis level regulation. Hindawi BioMed. Res. Int. **2017**, 12 (2017). Article ID 7432310

14. Nascu, I., Pistikopoulos, E.N.: Multiparametric model predictive control strategies of the hypnotic component in intravenous anesthesia. In: IEEE International Conference on Systems, Man, and Cybernetics (SMC), Budapest, Hungary, 9–12 October 2016

15. Nascu, I., Krieger, A., Ionescuan, C.M., Pistikopoulos, E.N.: Advanced model-based control studies for the induction and maintenance of intravenous anaesthesia. IEEE Trans. Biomed. Eng. **62**(3), 832–841 (2015)

16. Nascu, I., Oberdieckand, R., Pistikopoulos, E.N.: Offset-free explicit hybrid model predictive control of intravenous anaesthesia. In: IEEE International Conference on Systems, Man, and Cybernetics, Kowloon, China, 9–12 October 2015)

17. Zulfatman, H., Rahmat, M.F.: Application of self-tuning fuzzy PID controller on industrial hydraulic actuator using system identification approach. Int. J. Smart Sens. Intell. Syst. **2**(2) (2009)

18. Shi, S., Wang, D.: Study of the permanent magnetic drive speed control system based on single-output PID neural network. In: IEEE Fourth International Conference on Multimedia Information Networking and Security, pp. 262–265 (2012)

19. Chopra, V., Singla, S.K., Dewan, L.: Comparative analysis of tuning a PID controller using intelligent methods. Acta Polytechnica Hungarica **11**(8), 235–249 (2014)

A Novel Biogeography Inspired Trajectory-Following Controller for National Instrument Robot

Basma Jumaa Saleh$^{(\boxtimes)}$ [ORCID], Ali Talib Qasim al-Aqbi [ORCID],
and Ahmed Yousif Falih Saedi [ORCID]

Computer Engineering Department, Al-Mustansiriyah University, Baghdad, Iraq
eng.basmaj@gmail.com, alnser.Ali@gmail.com,
mr.a7med86@gmail.com

Abstract. This paper is devoted to the design of a trajectory-following control for a differentiation nonholonomic wheeled mobile robot. It suggests a kinematic nonlinear controller steer a National Instrument mobile robot. The suggested trajectory-following control structure includes two parts; the first part is a nonlinear feedback acceleration control equation based on adaptive sliding mode control that controls the mobile robot to follow the predetermined suitable path; the second part is an optimization algorithm, that is performed depending on the Mutated Harmony Search algorithm to tune the parameters of the controller to obtain the optimum trajectory. The simulation is achieved based on MATLAB R2017b and the results present that the kinematic nonlinear controller with MHS algorithm is more effective and robust than the original Harmony search learning algorithm; It is shown that the proposed scheme is robust to reduce the chattering problem because of adaptive control law of sliding mode controller; this is shown by the minimized tracking-following error to equal or less than (1 cm) and getting smoothness of the linear velocity less than (0.2 m/s), and all trajectory-following results with predetermined suitable are taken into account. Stability analysis of the suggested controller is proven using the Lyapunov method.

Keywords: Trajectory-following mobile robot · Sliding mode control
Kinematic nonlinear controller · National instrument
Harmony search algorithm

1 Introduction

Over the last few years, the interest in the machine learning and how it has been employed to help mobile robots in the navigation has been increased. Mobile robots have a lot of potential in the industry and in the house servant applications. Navigation control of wheeled mobile robots has been studied by many sources in the last decade ever after they are progressively used in an extensive range of applications. The motion of mobile robots must be modeled by mathematical calculation to estimate physical environment and run in delineated trajectories. At first, the research work was focused only on the kinematic model, assuming that there is optimum linear and angular

© Springer Nature Switzerland AG 2018
S. O. Al-mamory et al. (Eds.): NTICT 2018, CCIS 938, pp. 171–189, 2018.
https://doi.org/10.1007/978-3-030-01653-1_11

velocity following [1]. Mobile robots are mostly used in manufacturing, healthcare and medicine [2].

The basic estimation problems in wheeled mobile robot trajectory-following are still waiting to be addressed; then the implicit essence of the motivation for this paper is to generate optimum control parameters for the mobile robot, to follow the required path with small trajectory-following error, to beat unmolded kinematic disturbances, and to maintain the battery energy of the robot system. Traditionally, the problems of wheeled mobile robot controller have been processes by stabilizing point or by pre-determined suitable problems like a trajectory-following controller [3].

The navigation problem is the basic aim of a wheeled mobile robot in order that it requires a decision where to direct and that information is taken by an actual path. The problem of following may be classified into dynamic and kinematic trajectory-following control model. Kinematics controller targets are to reach angular and linear velocity as output to converge trajectory-following error to zero [4].

The modern survey of improvement nonholonomic control systems is illustrated in [5]. To the authors' knowledge, the trajectory-following control problem is one of research challenges for robot nonholonomic mobile systems in but has yet to be widely studied. From past researches, as an approximation for trajectory-following control, sliding mode control has been applied to the trajectory-following control of robots [6] and is lately receiving expansive interest of researches about nonholonomic control systems.

The feature of utilizing sliding mode control involves: changes in parameters is robust, rapid response and perfect transient execution but the chattering was the main challenge for its implementation [7]. Bloch and Drakunov [8] suggested a chained function for sliding mode tracking-following control problems. Guldner and Utkin [9], suggested a Lyapunov theory to define a set of given configuration for required trajectories, then given a trajectory-following feedback controller to minimizing error and reducing chattering.

In this work, we propose an adaptive optimum sliding mode controller for solving trajectory-following problems for a robot nonholonomic mobile system, then a required trajectory with asymptotic stability can be obtained using the suggested control.

In engineering problems, the optimization is to detect a solution that can minimize or maximize a cost function. These days, the stochastic method is more often used to solve the optimization problem [10]. Recently, the nature-inspired algorithm is proving its ability in solving numerical problems of optimization more suitably. These optimization ways are improved to solve complex problems, such as scheduling of flow shop [11], high-dimensional function optimization, reliability, and other engineering matters. Recently, many other algorithms have been used, like Artificial Bee Colony (ABC), Ant Colony Optimization (ACO), Particle Swarm Optimization (PSO), Harmony Search (HS) and Firefly Algorithm (FA).

Harmony Search (HS) is a relatively new heuristic optimization algorithm, and it was first developed by Geem et al. 2001 [10]. Since then, it has been used to solve different problems of optimization, including function optimization, water apportionment network, groundwater modeling, structural design, energy-salvage, and others.

2   Modeling of Differentiation Nonholonomic NI-Robot

In this research, we consider a differentiation two-wheeled drive mobile robots collected of two standard drive wheels on the same axis and a free driving castor wheel for stability. In order that robot positioning, trajectory-following, the linear velocity, angular velocity and kinematic model, as shown in Fig. 1. The variables for this mathematical kinematic model of a differentiation two-wheeled drive mobile robot are an actual positional and angle parameterization (x, y, θ), desired positional and angle parameterization (x_r, y_r, θ_r), radius of each wheel (R_a), centre of robot is (c) and length between two wheels (L_1), together with linear and angular velocities (V, Ω), respectively [12].

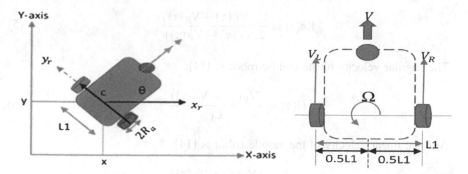

Fig. 1. Model of a differentiation mobile robot.

At each instant in time, the left and right wheels follow a trajectory as shown in Fig. 2 that moves around the instantaneous centre of curvature mobile robot (ICCM) with the same angular rate [13].

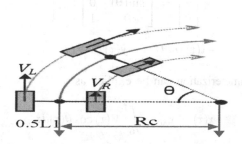

Fig. 2. The instantaneous centre of curvature.

$$\Omega(\tau) = \frac{d\theta(\tau)}{d\tau} \tag{1}$$

Thus:

$$V(\tau) = \Omega(\tau)Rc(\tau) \tag{2}$$

$$V_L(\tau) = \left(Rc(\tau) + \frac{L_1}{2}\right)\Omega(\tau) \tag{3}$$

$$V_R(\tau) = \left(Rc(\tau) - \frac{L_1}{2}\right)\Omega(\tau) \tag{4}$$

After solving Eqs. (3, 4), we found the instantaneous centre of curvature mobile robot (ICCM) path close to the centre point axis (c) is given as Eq. (5) [13].

$$Rc(\tau) = \frac{L_1(V_L(\tau) + V_R(\tau))}{2(V_L(\tau) - V_R(\tau))} \tag{5}$$

The angular velocity of the mobile robot is [14]:

$$\Omega(\tau) = \frac{(V_L(\tau) - V_R(\tau))}{L_1} \tag{6}$$

And the linear velocity of the mobile robot is [14]:

$$V(\tau) = \frac{(V_L(\tau) + V_R(\tau))}{2} \tag{7}$$

Therefore, the prediction kinematic model shown in Eq. (8). The constraint is non-holonomic, shown in Eq. (9) should be respected at rolling no spilling [15].

$$\begin{bmatrix} \dot{x} \\ \dot{y} \\ \dot{\theta} \end{bmatrix} = \begin{bmatrix} \cos(\theta) & 0 \\ \sin(\theta) & 0 \\ 0 & 1 \end{bmatrix} \begin{bmatrix} V \\ \Omega \end{bmatrix} \tag{8}$$

$$-\dot{x}(\tau)\sin\theta(\tau) + \dot{y}(\tau)\cos\theta(\tau) = 0 \tag{9}$$

The (x, y, θ) parameterization can be explicit as

$$x(\tau) = x_{00} + \int_0^\tau V(\tau)\cos\theta(\tau)d\tau \tag{10}$$

$$y(\tau) = y_{00} + \int_0^\tau V(\tau)\sin\theta(\tau)d\tau \tag{11}$$

$$\theta(\tau) = \theta_{00} + \int_0^\tau \Omega(\tau)d\tau \tag{12}$$

The discrete kinematic model by using Euler's theory with the instant time (L) and the sampling time (τ), can be expressed by Eqs. (13, 14, 15):

$$x(L) = 0.5[V_R(L) + V_L(L)]cos\theta(L)\Delta t + x(L-1) \tag{13}$$

$$y(L) = 0.5[V_R(L) + V_L(L)]sin\theta(L)\Delta t + y(L-1) \tag{14}$$

$$\theta(L) = \frac{1}{R_a}[V_R(L) + V_L(L)]\Delta t + \theta(L-1) \tag{15}$$

Finally, Fig. 3 show the configuration trajectory-following error $S_{ee} = [x_{ee}, y_{ee}, \theta_{ee}]^T$ can be presented by $S_{ee} = R * \mathfrak{E}$, where $\mathfrak{E} = [x_e, y_e, \theta_e]^T$ and (R) rotational matrix, which can be shown by the Eqs. (16, 17) [16]:

$$S_{ee} = \begin{bmatrix} x_{ee} \\ y_{ee} \\ \theta_{ee} \end{bmatrix} = \begin{bmatrix} cos(\theta(\tau)) & sin(\theta(\tau)) & 0 \\ -sin(\theta(\tau)) & cos(\theta(\tau)) & 0 \\ 0 & 0 & 1 \end{bmatrix} \begin{bmatrix} x_e \\ y_e \\ \theta_e \end{bmatrix} \tag{16}$$

$$\begin{bmatrix} x_e \\ y_e \\ \theta_e \end{bmatrix} = \begin{bmatrix} x_r - x \\ y_r - y \\ \theta_r - \theta \end{bmatrix} \tag{17}$$

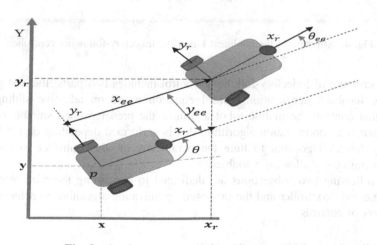

Fig. 3. Configuration error of mobile robot [12].

3   Trajectory-Following Controller

The proposed kinematic nonlinear controller is represented by the block diagram illustrated in Fig. 4. The mutated harmony search optimization algorithm which is used to optimum tuning parameters of adaptive sliding mode kinematic controller to minimize the trajectory-following error, generates the perfect velocity control actions (linear and angular velocity) and tracks the desired trajectory-followingcoordinates.

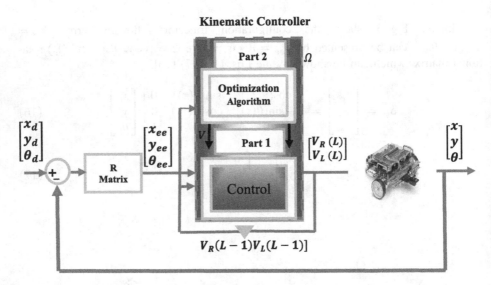

Fig. 4. Structure of the nonlinear kinematic trajectory-following controller.

The structure of trajectory-following control includes two parts; the first part is a nonlinear feedback acceleration control equation based on adaptive sliding mode control that controls the mobile robot to follow the predetermined suitable path; the second part is an optimization algorithm, that is performed depending on the Mutated Harmony Search algorithm to tune the parameters of the controller to obtain the optimum trajectory-following variables.

The following two subsections are dedicated to explaining the nonlinear sliding mode kinematic controller and the proposed optimization algorithm to achieve tuning parameters of controls.

3.1   Sliding Mode Kinematic Controller

The most commonly used design operations of SMC are given in the following subsections. Ordinarily, the process of sliding mode kinematic nonlinear control can be classified into two stages: (1) the choice of suitable switching surfaces such that if the system trajectory-following is bounded to lie upon it, then the control exhibited the required action; (2) the definition of a nonlinear control rule which is discontinuous on the ramified and is able to constrain the system trajectory to arrive at the ramified and stay on it. [17] These two stages are described below:

Design of Switching Function. The purpose of a tracking-following control is to make the parameterization trajectory-following error converge to zero for a qualitative desired trajectory as elapses of time. The kinematic configuration error Eq. (16) for mobile robot represents a nonlinear multiple-input system. For that reason, the design of switching function is a complex challenge. Depending on [18], to facilitate the problem, $x_{ee} = 0$ is selection as the first switching surface when $x_{ee} = 0$, the Lyapunov candidate function can be as

$$V = \frac{1}{2}y_{ee}^2 \qquad (18)$$

If $\theta_{ee} = -tan^{-1}(V_r, y_{ee})$, is selected as a switching candidate function, then

$$\dot{V} = y_{ee}\dot{y}_{ee} = y_{ee}(-x_{ee}\Omega + V_r sin\theta_{ee}) \qquad (19)$$

$$= -y_{ee}x_{ee}\Omega + y_{ee}V_d sin(-tan^{-1}(V_r, y_{ee})) \qquad (20)$$

$$= -y_{ee}x_{ee}\Omega - y_{ee}V_r sin(tan^{-1}(V_r, y_{ee})) \qquad (21)$$

Theorem 1: For any $x \in R$ and $|x| < \infty$ at $x = 0$.

$$\vartheta(x) = xsin(tan^{-1}x) \geq 0$$

Proof: Distinguishes the following three kind of situation discussion.
When $x = 0$, $\vartheta(0) = 0.x \in (0, +\infty), tan^{-1}x \in (0, +\frac{\pi}{2})$, *then*

$$sin(tan^{-1}x) \geq 0, \vartheta(x) > 0.x \in (-\infty, 0), \ tan^{-1}x \in \left(-\frac{\pi}{2}, 0\right), \ then$$

$$sin(tan^{-1}x) < 0, \vartheta(x) < 0.$$

From Theorem 1, can know
$y_{ee}V_r sin(-tan^{-1}(V_r, y_{ee})) \geq 0$, *when* $y_{ee}V_r = 0$, *then* $\dot{V} \leq 0$
It can obtain the conclusion, so long as convergence to zero and θ_{ee} convergence up to $tan^{-1}(V_r, y_{ee})$, then the system model convergence to zero. For that reason, according to this conclusion, the cut function is

$$S = \begin{bmatrix} S_1 \\ S_2 \end{bmatrix} = \begin{bmatrix} x_{ee} \\ \theta_{ee} + tan^{-1}(V_r, y_{ee}) \end{bmatrix} \qquad (22)$$

If $S_1 \rightarrow 0$, $S_2 \rightarrow 0$ and x_{ee} convergence into zero, and θ_{ee} convergence up to $-tan^{-1}(V_r, y_{ee})$, then, $y_{ee} \rightarrow 0, \theta_{ee} \rightarrow 0$.

Design of the Adaptive Control Rule. After the sliding surface is designed, we should design the nonlinear control rule that makes every state vector arrive at the sliding surface. To do so, design the nonlinear control rule with the next reaching condition [18]. Control law improved to reduce the chattering in Eq. (23) by using auto-tuning control swarm-based optimization algorithm to find and tune the optimum values of (δ_i) in addition $(\mathbb{K})$ from [19].

$$\dot{\mathbb{S}} = \mathbb{K} \, sat \, (\mathbb{S}) \tag{23}$$

Then use continuous function substitutes for the sign function

$$\dot{\mathbb{S}}_i = -\mathbb{K}_i \frac{\mathbb{S}_i}{|\mathbb{S}_i| + \delta_i} i = 1,2 \tag{24}$$

Set $\boldsymbol{\epsilon} = tan^{-1}(V_r, y_{ee})$, from Eqs. (8) and (22), yield,

$$\dot{\mathbb{S}} = \begin{bmatrix} \dot{\mathbb{S}}_1 \\ \dot{\mathbb{S}}_2 \end{bmatrix} = \begin{bmatrix} \mathbb{K}_1 \frac{\mathbb{S}_1}{|\mathbb{S}_1| + \delta_1} \\ \mathbb{K}_2 \frac{\mathbb{S}_2}{|\mathbb{S}_2| + \delta_2} \end{bmatrix} \tag{25}$$

$$\dot{\mathbb{S}} = \begin{bmatrix} \dot{x}_{ee} \\ \dot{\theta}_{ee} + \frac{\partial \epsilon}{\partial v_r} \dot{V_r} + \frac{\partial \epsilon}{\partial y_{ee}} \dot{y}_{ee} \end{bmatrix} \tag{26}$$

$$= \begin{bmatrix} V_r \cos(\theta_{ee}) + y_{ee}\Omega - V \\ \Omega_r - \Omega + \frac{\partial \epsilon}{\partial V_d} \dot{V_r} + \frac{\partial \epsilon}{\partial y_{ee}} V_r \sin(\theta_{ee}) - x_{ee}\Omega \end{bmatrix} \tag{27}$$

Rearrange Eq. (28), we have

$$V_g = \begin{bmatrix} V \\ \Omega \end{bmatrix} = \begin{bmatrix} y_{ee}\Omega + V_r \cos(\theta_{ee}) + \mathbb{K}_1 \frac{\mathbb{S}_1}{|\mathbb{S}_1| + \delta_1} \\ \dfrac{\Omega_r + \frac{\partial \epsilon}{\partial v_r} \dot{V_r} + \frac{\partial \epsilon}{\partial y_{ee}} V_r \sin(\theta_{ee}) + \mathbb{K}_2 \frac{\mathbb{S}_2}{|\mathbb{S}_2| + \delta_2}}{1 + \frac{\partial \epsilon}{\partial y_{ee}} x_{ee}} \end{bmatrix} \tag{28}$$

Where $\frac{\partial \epsilon}{\partial y_{ee}} = \frac{V_r}{1 + (V_r y_{ee})^2}$, $\frac{\partial \epsilon}{\partial V_r} = \frac{y_{ee}}{1 + (V_r y_{ee})^2}$

Finally, the time-varying nonlinear control rule depending on sliding mode kinematic control can be described by Eq. (29), from [20]:

$$V_g = \begin{bmatrix} V_R \\ V_L \end{bmatrix} = \begin{bmatrix} V + \frac{L1}{2}\Omega \\ V - \frac{L1}{2}\Omega \end{bmatrix}. \tag{29}$$

3.2 Swarm-Based Optimization Algorithms

The stochastic algorithm which uses multiple agents (solutions) to move through the search space in the process of solving an optimization problem is known as the optimization algorithm. Some of these effective stochastic techniques that mimic the behaviours of certain animals or insects (birds, ants, bees, flies and even germs) are called Nature-Inspired Algorithm used for auto-tuning controllers. Two of these techniques are discussed here:

Harmony Search Algorithm. Harmony Search (HS) algorithm is a search heuristic based upon the improvisation process of jazz musicians by Geem et al. [10]. In jazz music, the different musicians try to adjust their pitches, such that the overall harmonies are optimized due to aesthetic objectives, so it is based on the process that musicians carry out when composing music [21]. The parameters of the HS are given [22].

1. Size of Harmony Memory (SHM) (i.e. number of member vectors in memory of harmony).
2. Considering Rate of Harmony Memory (CRHM), as CRHM $\in$ [0,1].
3. Pitch Adjustment Rate (PAR), as PAR $\in$ [0,1] and Pitch Adjustment Bandwidth (bw).
4. Stopping Condition (i.e. number of improvisation (NJ)). More explanations of these parameters are in the next steps:

- Initialization:

The Harmony Memory (HM) is a matrix of members with a size of SHM, where each harmony memory vector presents one member as able to be shown in Eq. (30). In this section, the members are manufactured randomly and reordered in an inverted order to HM, depending on fitness function [22].

$$HM = \left\langle \begin{matrix} a_1'^1 & a_2'^1 & a_{sm}'^1 & fitness(a'^1) \\ \vdots & \vdots & \vdots & \vdots \\ a_1'^{HMS} & a_2'^{HMS} & a_{sm}'^{HMS} & fitness(a'^{HMS}) \end{matrix} \right\rangle \qquad (30)$$

- Improvise new harmony:

This point is the substance of the HS algorithm and the fundamental that has been structuring this algorithm. In this point, the HS creates (improvises) a new harmony vector, $a' = (a_1', a_2', a_3', \ldots, a_{sm}')$. It can be explained according to the discussion on the player improvisation process [22]. There are three feasible options for a player in the music improvisation procedure: (1) play several pitches that are the same with the CRHM; (2) play some pitches like a known piece; alternatively, (3) improvise new pitches, depending on three operators: consideration of memory; adjustment of pitch; on the other hand, consideration is randomly [23]. In the consideration of memory, the values to the new harmony vector are randomly patrimonial from the historic values saved in HM with a probability of CRHM. Thus, the value of decision adjustable (a'_1) is selected from $(a'_1, a'_1, a'_1, \ldots, a'_{sm})$ that is saved in HM. The next decision adjustable (a'_2) is selected from $(a'_2, a'_2, a'_2, \ldots, a'_{sm})$, and the other decisions adjustable, $(a'_3, a'_4, a'_5, \ldots)$ are selected sequentially in the same method with the probability of CRHM $\in [0, 1]$. In most cases, (CRHM = 0.7–0.95). The work of HM is like the musician utilizing his or her memory to produce a perfect tune. That accumulative stage assures that excellent harmonies are regarded as the members vectors of New Harmony [22] So, choose a value from HM according to Eq. (31) [23].

$$a'_{new} = a'_b \text{ where } b \in (1, 2, \ldots, HMS) \tag{31}$$

• Pitch adjusting:
The pitch in the second operator needs to be adjusted slightly by Eq. (32) [10].

$$a'_{new} = a'_{old} + bw(2\epsilon - 1) \tag{32}$$

Where ϵ is a random number in [0, 1] and a_{old} is the present pitch. Here, bw is the pitch adjustment bandwidth. Parameter PAR should also be appropriately set. If PAR is very close to 1, after that the solution is always updating, and HS is complex to stable. If it is next to 0, then little change is made, and HS may be precocious [23]. So, here we set PAR = 0.1–0.7.

• Randomization:
Output of that, where the harmony elements are not selected from HM, depending on the CRHM probability exam, they are randomly selected depending on their predetermined range, as given by Eq. (33) [22]:

$$a'_{new} = min + rand * (max - min) \tag{33}$$

Where $rand \in [0, 1]$, max is a maximum observable value, and min is a minimum observable value.

Mutated Harmony Search Algorithm. Harmony search (HS) algorithm (Geem et al. 2001) [10] and biogeography-based optimization (BBO) (Simon 2008) [24] are two relatively new EAs. This research proposed a new method for improving harmony search algorithm by using mutation from biogeography-based optimization. HS imitates the improvisation process of musicians by creating new harmonies (solutions) based on a harmony memory (HM) and pitch-adjustment operations. On the other hand, mutation operator plays an important role to explore the search space, retain the diversity of individuals and break away local optimums. Finally, the proposed method showed effectiveness on a wide variety of optimization problems and enriched harmony diversity. The mutation operator can be applied by replacing Eq. (33) by Eq. (34). On the other hand, new harmonies (solutions) generated based on mutation instead of randomly by Eq. (34) and the pseudocode at Algorithm 1.

$$a_{new} = floor(minpar + (maxpar - minpar + 1) * (rand)) \qquad (34)$$

Algorithm 1. The summarization of the computation procedure of hybrid MHS algorithm

Step 1: *Initialize the HM.*

Step 2: *Evaluate the fitness function f (a'), $(a'_1, a'_2, ..., a'_{sm})$.*

Step 3:*define(CRHM), (SHM), (PAR)*

Step 4:*define maximum number of iteration, minpar and maxpar.*

Step 5:*While the halting criteria is not satisfied do*

        For d=1: sm do

If rand <= CRHM then // memory consideration Choose the value by Eq.(32)

    If rand <= PAR then // pitch adjustment to adjust the value by Eq.(33)

End if,

Else   // random selection

Select random value by Eq. (34).

End if,

End for d

Step 6:*Update the HM as $a'_w = a'_{new}$, if $(a'_{new}) < (a'_w)$..*

Step 7:*End*

3.3   Design of Kinematic Nonlinear Controller Based on MHS Algorithm

This work introduced a mutated harmony search algorithm to find the optimum parameter of control. The searching procedure of the suggested algorithm is described as follows:

1. Initial searching elements of each harmony are created randomly within the determined range. Note that the dimension of search space consists of all the control parameters needed in the adaptive sliding mode nonlinear control as described in Fig. 4.

2. The fitness function for each harmony can be calculated by the mean square error theory as Eq. (35) from [20]:

$$(MSE) = \frac{1}{2} \sum_{p=1}^{pop} (x_r(L)^p - x(L)^p)^2 + (y_r(L)^p - y(L)^p)^2 + (\theta_r(L)^p \theta(L)^p)^2 \qquad (35)$$

3. Set step 3 to step 7 of the MHS algorithm pseudo code to get the perfect value of controller's gain by minimizing trajectory-following error.

4. If the epochs number reach at max, then exit, else, go to step 2.

4 Simulation

In The aim of the simulations is to examine the effectiveness and performance of the proposed adaptive sliding mode kinematic nonlinear control depending on original harmony search and MHS optimization algorithms to a differentiation two-wheeled drive mobile robot by programming using MATLAB environment.

The resulting of a differentiation two-wheeled drive mobile robot trajectory-following, obtained by the suggested kinematic nonlinear control is involving trajectory-following, tracking-following error, the linear velocity of right and left wheel, linear and angular velocity of the mobile robot and (MSE). The execution of work can be simulated by the following table of the setting parameter (Harmony elements is equal to 3, NI-robot length is equal to 0.36 m and NI-robot radius is equal to 0.05 m).

Case Study: (Lemniscates Trajectory). The desired trajectory-following lemniscates, can be expressed by the following equations:

$$x_r(\tau) = -0.5 * \sin\left(\frac{2\pi\tau}{30}\right), \quad y_r(\tau) = 0.5 * \sin\left(\frac{2\pi\tau}{20}\right)$$

$$\theta_r(\tau) = 2 \tan^{-1}\left(\frac{\Delta y_r(\tau)}{\sqrt{(\Delta x_r(\tau))^2 - (\Delta y_r(\tau))^2} + \Delta x_r(\tau)}\right)$$

- The initial desired parameterization of robot at $q_r = \left[0.1, 0, \frac{pi}{2}\right]$ and the initial actual parameterization of robot at $q = [0, 0, 0]$, then the simulation of trajectory-following lemniscates of the NI mobile robot using adaptive sliding mode kinematic control based on original HS learning algorithm is viewed in Fig. 5. And the simulation of trajectory-following lemniscates of the NI mobile robot using adaptive sliding mode kinematic controller based on MHS learning algorithm is viewed in Fig. 6.
- The left and right velocity of adaptive sliding mode kinematic control based on original HS algorithm is shown in Fig. 7, and for adaptive sliding mode kinematic controller based on MHS learning is shown in Fig. 8.
- The performance index MSE for parameterization trajectory-tracking during 10 epochs are described in Fig. 9 which is (1.071) for the proposed control based on original HS learning algorithm and Fig. 10 which is (1.070) for the proposed control based on MHS in the simulation case study, which has perfect parameterization trajectory-following performance and it had the ability of generating suitable and smooth velocity without spike.
- Figure 11 proves the average of linear velocity (0.01 m/s) and the top peak of the angular velocity (± 0.01 rad/s) of the NI robot for the proposed control based on original HS learning algorithm and Fig. 12 proves the average of linear velocity (0.185 m/s) and the top peak of the angular velocity (± 1.06 rad/s) of the NI robot for proposed control based on MHS.

- Instantaneous suggested control parameters are shown in Fig. 13 which is tuned by using mutated harmony search.
- The performance described in (Figs. 14, 15 and 16) for the proposed control based MHS algorithm is obvious by showing the convergence of parameterization trajectory-following errors with small value.
- Figure 17 shows the sliding surfaces with chattering (original harmony search) and Fig. 18 shows the sliding surfaces without chattering (mutated harmony search) applied on control.

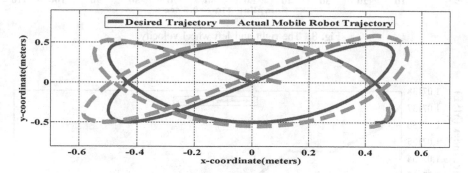

Fig. 5. Actual and desired Lemniscates trajectory-following.

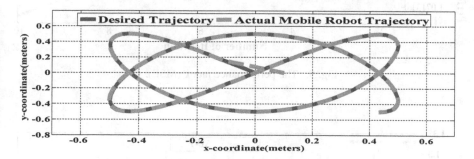

Fig. 6. Actual and desired Lemniscates trajectory-following.

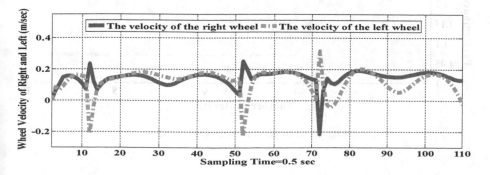

Fig. 7. The right and left wheel velocity.

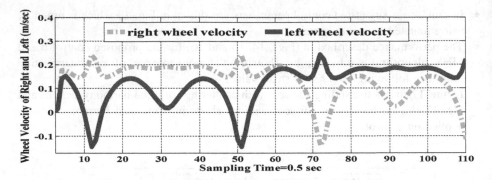

Fig. 8. The right and left wheel velocity.

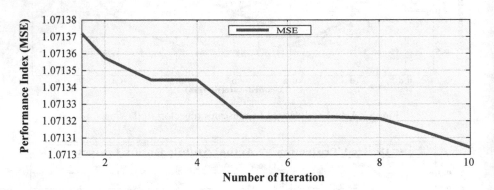

Fig. 9. The performance index (MSE).

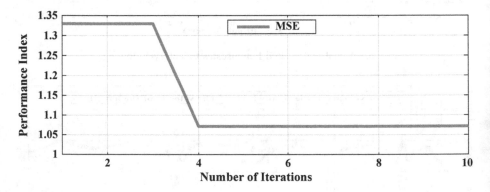

Fig. 10. The performance index (MSE).

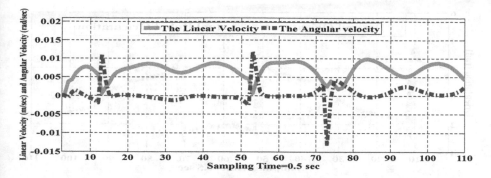

Fig. 11. The linear and angular velocity.

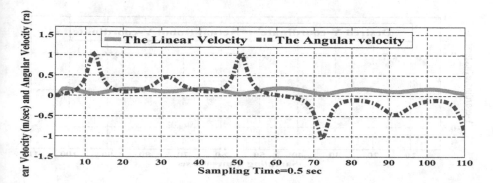

Fig. 12. The linear and angular velocity

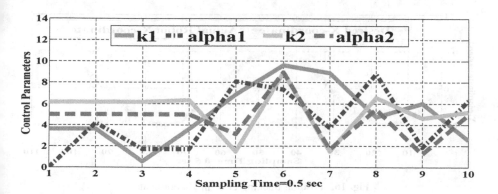

Fig. 13. Control parameters.

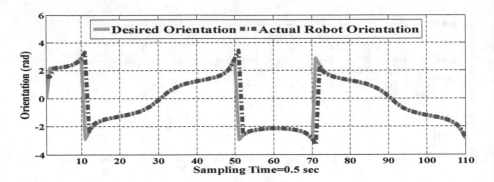

Fig. 14. Orientation tracking error

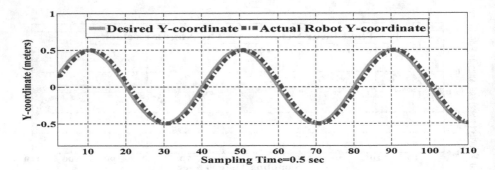

Fig. 15. Position tracking error in Y-coordinate.

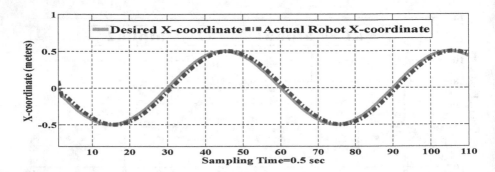

Fig. 16. Position tracking error in X-coordinate.

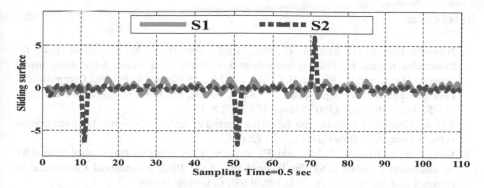

Fig. 17. Sliding surfaces.

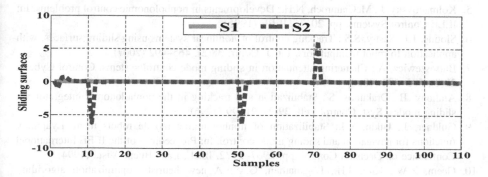

Fig. 18. Sliding surfaces

5 Conclusion

The trajectory-following nonlinear kinematic control based on sliding mode theory with two optimization algorithms for the differentiation nonholonomic two wheeled drive mobile robot have been presented in this work. The suggested controller examined by MATLAB (R2017b) set on National Instrument mobile robot (NI Mobile Robot). The simulation results show evidently the capability of robustness and adaptation of the suggested sliding mode kinematic controller based on mutation harmony search algorithm which has excellent sliding surfaces and control gains without chattering, optimum trajectory-following performance and it has the capability of generating more suitable and smooth velocities than the controller based on original harmony search because the MHS algorithm has capability to get optimum controller parameters.

References

1. Narendra Kumar, D., Samalla, H., Rao, Ch.J., Swamy Naidu, Y., Alfoni Jose, K., Manmadha Kumar, B.: Position and orientation control of a mobile robot using neural networks. In: Jain, L.C., Behera, H.S., Mandal, J.K., Mohapatra, D.P. (eds.) Computational Intelligence in Data Mining - Volume 2. SIST, vol. 32, pp. 123–131. Springer, New Delhi (2015). https://doi.org/10.1007/978-81-322-2208-8_13
2. Seo, K., Lee, J.S.: Kinematic path-following control of mobile robot under bounded angular velocity error. Adv. Robot. **20**(1), 1–23 (2006)
3. Kanayama, Y., Kimura, Y., Miyazaki, F., Noguchi, T.: A stable tracking control method for an autonomous mobile robot. In: Proceedings of the IEEE International Conference on Robotics and Automation, pp. 384–389. IEEE, Cincinnati (1990)
4. Zain, A.A., Daobo, W., Muhammad, S., Wanyue, J., Muhammad, S.H.: Trajectory tracking of a nonholonomic wheeleed mobile robot using hybrid controller. Int. J. Model. Optim. **6** (3), 136–141 (2016)
5. Kolmanovsky, I., McClamroch, N.H.: Developments in nonholonomic control problems. In: IEEE Control Systems, pp. 20–36. IEEE (1995)
6. Slotine, J.J., Sastry, S.S.: Tracking control of nonlinear systems using sliding surfaces, with application to robot manipulators. Int. J. Contr. **38**(2), 465–492 (2007)
7. Bartoszewicz, A.: Chattering attenuation in sliding mode control systems. Control Cybern. **29**(2), 585–594 (2000)
8. Anthony, B., Drakunov, S.: Stabilization and tracking in the nonholonomic integrator via sliding modes. Syst. Control Lett. **29**(2), 91–99 (1996)
9. Guldner, J., Utkin, V.I.: Stabilization of nonholonomic mobile robots using Lyapunov functions for navigation and sliding mode control. In: Proceedings of the IEEE International Conference on Decision Control, pp. 2967–2972. IEEE, Lake Buena Vista (1994)
10. Geem, Z.W., Kim, J.H., Loganathan, G.V.: A new heuristic optimization algorithm: harmony search. SAGE J. **76**(2), 60–68 (2001)
11. Adithyan, T., Vasudha, S., Gururaj, B., Chandrasegar, T.: Nature inspired algorithm. In: International Conference on Trends in Electronics and Informatics (ICEI), pp. 1131–1134. IEEE, Tirunelveli (2017)
12. Nizar, H., Basma, J.: Trajectory Tracking Controllers for Mobile Robot: Modeling, Design and Optimization. Lambert Academic Publishing, Saarbrücken (2016)
13. Al-Araji, A.: Design of a cognitive neural predictive controller for mobile robot. Ph.D. thesis, Brunel University, UK (2012)
14. Nizar, H., Basma, J.: Design of a kinematic neural controller for mobile robots based on enhanced hybrid firefly-artificial bee colony algorithm. Al-Khwarizmi Eng. J. **12**(1), 45–60 (2016)
15. Ye, J.: Adaptive control of nonlinear PID-based analogue neural network for a nonholonomic mobile robot. Neurocomputing **71**(7), 1561–1565 (2008)
16. Yousif, Z., Hedley, J., Bicker, R.: Design of an adaptive neural kinematic controller for a national instrument mobile robot system. In: IEEE International Conference on Control System, Computing and Engineering, pp. 623–628. IEEE, Penang (2012)
17. Jun, K.L., Yoon, H., Jin, B.: Sliding mode tracking control of mobile robots with approach angle in cartesian coordinates. Int. J. Control Autom. Syst. **13**(3), 718–724 (2015)
18. Chen, M.-L., Ko, Y.-H.: Wang, J-R: Slider controller design for two-wheeled mobile robot scheme. J. Chung Cheng Inst. Technol. **40**(2), 113–120 (2011)
19. Lee, J.H., Lin, C., Lim, H., Lee, J.M.: Sliding mode control for trajectory tracking of mobile robot in the RFID sensor space. Int. J. Control Autom. Syst. **7**(3), 429–435 (2009)

20. Al-Araji, A.S.: Development of kinematic path-tracking controller design for real mobile robot via back-stepping slice genetic robust algorithm technique. Arab. J. Sci. Eng. **39**(12), 8825–8835 (2014)
21. Amaya, I., Cruz, J., Correa, R.: Harmony search algorithm: a variant with self-regulated Fretwidth. Appl. Math. Comput. **9**(266), 1127–1152 (2015)
22. Moh'd, A.O., Mandava, R.: The variants of the harmony search algorithm: an overview. Artif. Intell. Rev. **36**(1), 49–68 (2011)
23. Wang, G., Guo, L.: A novel hybrid bat algorithm with harmony search for global numerical optimization. J. Appl. Math. 1–21 (2013)
24. Simon, D.: Biogeography-based optimization. Evol. Comput. IEEE Trans. Evol. Comput. **12** (6), 702–713 (2008)

Communication Applications

Modulation Mapping Influence in Coherent Optical OFDM System for Long Haul Transmission

Liqaa A. Al-Hashime[1]($\boxtimes$), Ghaida A. Al-Suhail[1]($\boxtimes$),
and Sinan M. Abdul Satar[2]($\boxtimes$)

[1] Department of Electrical and Computer Engineering,
University of Basrah, Basrah, Iraq
newliqas@gmail.com, ghaida_alsuhail@yahoo.com
[2] Department of Electrical and Electronic Engineering,
University of Technology, Baghdad, Iraq
sinansma@yahoo.com

Abstract. There is no doubt that Coherent Optical Orthogonal Frequency Division Multiplexing (CO-OFDM) is considered as an attractive modulation format. The previously mentioned term has received a great deal of attention in recent years within the long haul fiber-optic transmission community. However, the benefit of optical OFDM is concerned with indefinite quantity of eliminating Inter Symbol Interference (ISI). Such a quantity is caused due to Chromatic Dispersion (CD) and polarization-mode dispersion (PMD).

Consequently, CO-OFDM exhibits a robust and spectrally efficient modulation scheme for the succeeding generation of the high-speed optical communications. In this paper, the CO-OFDM system is analyzed according to a couple of modulation mapping orders which take into account the effect of continuous wave (CW) laser line width. Interestingly, this paper deals with the performance investigation of the system design in terms of transmission distance, Bit Error Rate (BER), Quality Factor (QF) and Error Vector Magnitude (EVM). It has been noticed that the system resistance against high Bit Error Rate is obtained when two certain mapping orders of 4QAM and 16QAM are considered at bit rate 10 Gbps. As a result, it has been found that the system can achieve up to 950 km and 660 km distances respectively, by using VPI simulator compared to the existing systems.

Keywords: Coherent Optical OFDM (Co-OFDM)
Orthogonal Frequency Division Multiplexing (OFDM)
Quadrature amplitude modulation (QAM) · Quality Factor (QF)
Error Vector Magnitude (EVM) · Bit Error Rate (BER)

1 Introduction

As is well known, OFDM has been extremely used in different telecommunication applications. This happens due to its high spectral proficiency and easy hardware implementation. It is worthwhile noting that such OFDM has been suggested by many authors as a promising candidate for the following generation of long-haul transmission and high data proportion in optical communication systems [1]. Given that Optical OFDM has been

© Springer Nature Switzerland AG 2018
S. O. Al-mamory et al. (Eds.): NTICT 2018, CCIS 938, pp. 193–209, 2018.
https://doi.org/10.1007/978-3-030-01653-1_12

considered as a new technique in optical networks in order to face the CD in fiber. Additionally, it is used to compensate electrically PMD in single-mode fiber (SMF) systems by using digital signal processing (DSP) at the recipient [1, 2]. Underlying the applications, OOFDM is principally classified into DirectDetectionoptical OFDM (DDO-OFDM) and Coherent Detection OFDM (CO-OFDM). Evidently, one photodiode is used in DDO-OFDM system, whereas in CO-OFDM the principle of optical combination is employed with Local Oscillator (LO) in optical coherent receiver consequently [2, 3].

Above all, the grander performance of CO-OFDM makes it apt for long-term networks in terms of its dispersion lenience and the Optical Signal to Noise Ratio (OSNR). Therefore, the aim of CO-OFDM is to realize high spectral efficiency in addition to receiving sensitivity in optical networks. A further important implication is that once the modulation technique of OFDM is combined with coherent detection, the advantages brought by these two powerful techniques are considered as unit multifold [1]: (1) high spectral efficiency; (2) sturdy to CD and PMD; (3) high receiver sensitivity; (4) Dispersion Compensation Modules (DCM); (5) less DSP complexity; (6) less oversampling factor; and more flexibility in spectral shaping and matched filtering. Likewise, CO-OFDM was initially suggested for the purpose of combatting CD in fiber [1]; and extended to be resilient to fiber PMD [4]. It was able to report the first experiment for 1000 km SSMF transmission at 8 Gb/s [5]. Moreover, it had quickly reported for about 4160 km SSMF transmission at 20 Gb/s [6]. Conversely, it has been noticed that CO-OFDM requires high complexity in the transceiver design in comparison with DDO-OFDM. Furthermore, as a result of its sensitivity to frequent offset and phase noise (PN), it has been illustrated that laser sources with very narrow line width are required at the transmitter and receiver sides [7]. The PN characteristics in DDO-OFDM systems are found to be quite different from that in CO-OFDM [8].

Nevertheless, to address the optical transmission impairments, further studies have reported that CO-OFDM system suffers mainly from linear and nonlinear impairments when using higher order of M-PSK or M-QAM, where (M) is the order of mapping [9]. Even though, coherent transmission system that is utilizing DSP permits the compensation of system feebleness. Examples of the previously mentioned case are CD, PMD, Laser Phase Noise (LPN), and Fiber Nonlinearities (FNLs) [10]. On the other hand, there are many other means to utilize the signal processing in order to reduce the impacts of CD, LPN and FNLs. Clearly, some of these methods have been suggested in the system receiver like time domain equalizer, frequency domain equalizer [11, 12] and [13]. Another way is also devoted to reduce LPN; where a part of sub carriers is not isolated to hold data but to do channel estimation, phase and frequency compensating in the receiver as in [7]. Additionally, some DSP methods are applied in the system transmitter. It has been illustrated that they actually depend on minimizing the Peak to Average Power Ratio (PAPR). A significant reduction can be captured when using one scheme of block coding techniques, Selective Mapping (SLM) [14], Partial Transmit Sequence (PTS) as in [15], non-linear transforms [16] and so on. Moreover, using MPSK or MQAM systems with DSP based dispersion compensation leads to strong influence of laser phase noise and this may need to use equalization enhanced phase noise [7, 13]. To put it another way, a variety of modulation schemes have a direct influence on the OSNR, optical fiber CD tolerance, non-linear effects and so on. To that end, it is essential to consider other design parameters and balance among them. They should also be chosen from the transmission distance, transmission capacity, and spectral efficiency. Furthermore, they required BER such as in [17] the performance of 16(D)APSK in CO-OFDM with and without differential encoding has proved more advantage than 16QAM for a single channel up to 1000 km. Also, the DCF in post-compensation scheme in QPSK is used when the input laser power is 5 dBm for 100 km [18]. The direct-detection and QAM-coherent detection systems are also studied using Optisystem, the latter one has a better performance in Q-factor and BER for a single and 4 channels [19].

The major objective is to investigate the CO-OFDM system model for a couple modulations of 4QAM and 16QAM by using VPI in comparison with the existing systems. It may be assumed that the analytical performance of the system in choosing a proper modulation mapping order and the predefined line width of the CW laser. Accordingly, the simulation results explained system performance in terms of the received signal. The paper is organized as follows. Section 2 gives a detailed description of CO-OFDM system. The third section illustrates the main performance quality metrics that are used in optical systems; whereas the simulation of system model by using VPI and co-simulation of Matlab to illustrate the system performance is described in the fourth section. Section 5 outlines the conclusions, in addition to future work.

2 CO-OFDM Description

Generally speaking, the thought behind multi-carrier Modulation (MCM) strategies, for example OFDM, is that one high data proportion stream is divided into different lower data rate streams. It has been said that these streams can be transmitted in parallel, offering longer symbol periods than single bearer procedures. Moreover, these parallel streams are ordinarily adjusted in OFDM by using one type of mapping. Good examples could be, BPSK, QPSK and QAM [20].

OFDM's, compact arrangement of subcarriers, which are orthogonal to each other in the frequency domain guarantees high spectral efficiency. Additionally, it is considered the fundamental explanation for OFDM use in current radio norms. Be that as it may, as expressed beforehand, what makes OFDM especially alluring for use in optical communications is its intrinsic resilience to CD [2]. The optical OFDM systems can be divided into two types according to the detection design are as DDO-OFDM and CO-OFDM. Regardless of the low cost benefits accrued in direct detection, coherent detection-based systems represent the best performance in receiver sensitivity, spectral efficiency and robustness against polarization dispersion, although coherent detection-based systems demand the highest complexity in the transmitter design. The CO-OFDM system can be divided into five parts: Radio Frequency (RF) OFDM transmitter, Optical Transmitter, Optical Channel, Optical Receiver and RF receiver. Figure 1 shows the basic building of Optical Homodyne OFDM system.

2.1 OFDM Transmitter

There is no doubt that in the transmitter, there are many steps in order to obtain an OFDM signal as shown in Fig. 1(b). The input Pseudo Random Bit Sequence (PRBS) is converted from a serial to parallel then being entered into the Mapping process in order to convert bits into symbols. Likewise, the inverse fast Fourier transformer

(IFFT) algorithm is executed to change signal from frequency domain to time domain. To put this in other words, when the signal is transformed from parallel to serial, the Cyclic Prefix (CP) is inserted in order to overcome the channel dispersion induced ISI and Inter-Carrier Interference (ICI) [2]. At the end, the signal's components go through a Digital to Analog Converter (DAC). Then, RF base band OFDM signal is obtained and can be mathematically expressed as [2].

$$S_b(t) = \sum_{i=-\infty}^{+\infty} \sum_{k=-\frac{Nsc}{2}+1}^{\frac{Nsc}{2}} S_{ki} \prod(t - T_s)e^{j2\pi f_k(t-iT_s)} \tag{1}$$

Where:

$$f_k = \frac{k-1}{t_s} \tag{2}$$

$$\prod(t) = \begin{cases} 1, & -\Delta G < t \le t_s \\ 0, & t \le -\Delta G, t > t_s \end{cases} \tag{3}$$

Where Nsc is the number of subcarrier, Ski is the symbol of information of Kth subcarrier. While, fk is the frequency of subcarrier, Ts is the symbol period. Whereas, ts is the observation period and ΔG is the guard length. In current system, the modulation schemes are considered as QAM which encodes nbit/symbol in rectangular constellation for amplitude and phase depending on the order of modulation. In order to eliminate the ISI between OFDM symbols, a guard symbol is used and should satisfy the following condition:

$$GI \ge (C\,|D_t|N_{sc})/(f^2 t_s) \tag{4}$$

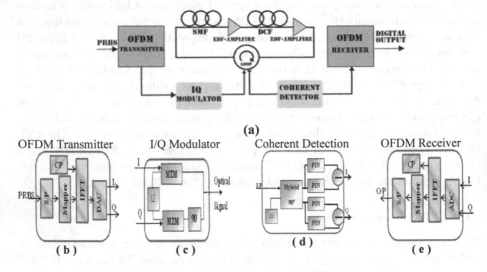

(a)

OFDM Transmitter   I/Q Modulator   Coherent Detection   OFDM Receiver

(b)   (c)   (d)   (e)

Fig. 1. Basic diagram of homodyne CO-OFDM system

2.2   IQ Modulator

The IQ modulator, which is shown in Fig. 1(c), consists of CW laser with two Mach-Zehnder modulators (MZMs). They are used in order to up convert the real/imaginary components of the signal, from the RF domain to the Optical domain. Furthermore, there is a phase difference between the two branches of 90°. Both signals from two branches are combined together to get the optical modulated signal. It has been illustrated that the IQ modulator has a main advantage. Such an advantage can be seen in the immunity against the rotation of the transmitted symbols (it adds a dimension to the signal) [21]. Moreover, to describe the output of the IQ modulator, Eq. 5 expresses the sum of the two sine waves but one of them is described with phase shift (90°). The first term of the equation is the real part and the second is the imaginary part.

$$E_o(t) = E_0 \left(\sin\left(\frac{s(t) * \pi}{V_\pi * 2}\right) + j\sin\left(\frac{s(t) * \pi}{V_\pi * 2}\right) \right) e^{jwt} \tag{5}$$

Basically, the signal is passed through the Optical Channel which consists of three parts. The first part is SMF, which is used to transfer the optical signal from the transmitter to the receiver. Whereas the second part is the post-DCF, which is used to compensate the dispersion and fiber nonlinearity in the optical channel. In this technique, DCF fiber is used after standard SMF, which is considered as a special kind of fiber that has a very large negative dispersion. Typically, DCF dispersion can be observed in the range of −80 ps/nm/km. A good example is a 20 km length of which can compensate for the dispersion in a 100 km length of fiber [22, 23]. Finally, the third part is Erbium-Doped Fiber Amplifier (EDFA), which is an optical device that is used to amplify the intensity of optical signals being carried through the optical channel. The signal out from the channel equals:

$$E_r(t) = E_s(t) * h(t) + noise_{ch}(t) \tag{6}$$

To put it another way, where Er received signal, Es sent signal, h(t) channel impulse response and noise of the channel.

2.3   Coherent Detection

The fourth part is the optical Receiver which is consisted of homodyne detection with a local oscillator as shown in Fig. 1(d) [2]. To convert the optical signal to RF signal. There are two types of coherent receiver, homodyne and heterodyne detection correspondingly. They can down-change an optical signal to a baseband electrical signal. In Fig. 1(d), the Coherent Homodyne Detection scheme is considered. This is similar to the Heterodyne Detection scheme except that the LO has the same frequency as the received optical carrier [24]. The 90° optical hybrids will generate a 90° phase shift among I and Q components and 180° phase shift among balanced detection. The outputs of the receiver are E1, E2, E3 and E4 from the inputs Er and Eq.

$$E_1 = \frac{1}{2}\left(E_r + E_q\right) \tag{7}$$

$$E_2 = \frac{1}{2}\left(E_r - E_q\right) \tag{8}$$

$$E_3 = \frac{1}{2}\left(E_r + jE_q\right) \tag{9}$$

$$E_4 = \frac{1}{2}\left(E_r - jE_q\right) \tag{10}$$

Where Er is the received signal and Eq is the LO signal. The complex detected signal comprising in-phase and quadrature phase components which is the output of coherent detector that is feeded to the OFDM receiver can be written as:

$$I(t) = I(t) + jQ(t) = 2E_r E_q^* \tag{11}$$

2.4   OFDM Receiver

In the receiver side, the signal would entry OFDM receiver which is shown in Fig. 1(e). The phase (I) and quadrature (Q) segments of the got signal would bear the reverse methodology applied at the transmitter side. Also, initializing the Analog to Digital Converter (ADC) would be applied. After that, associate OFDM is required to remove CP. Then the signal is prepared to be introduced within the fast Fourier transformer (FFT) algorithm which is converted from time domain to frequency domain. Next, the signal is converted from parallel to serial after doing de-mapping process so as to get the original signal.

3   Quality Performance Metrics

In this section, some outstanding performance criteria are being briefly outlined which can be used to evaluate the optical OFDM communication systems as follows:

3.1   Quality Factor and Optical Signal to Noise Ratio

OSNR is acquired as the proportion of the signal power to the noise power. The principle hotspot for its corruption is noise embedded by optical enhancers. Moreover, QF is another vital parameter that is utilized as a part of this paper for the investigation of simulations. The QF can be written as takes after:

$$Q(dB) = 20 * \log 10\left[\sqrt{2}\mathrm{erfc}^{-1}(2BER)\right] \tag{12}$$

Basically, QF determines the minimum required OSNR to get a specific estimation of BER. The mathematical relation amongst QF and BER is given by:

$$BER = 0.5 \text{ercf} \left[Q_{\text{linear}} / \sqrt{2} \right] \tag{13}$$

In general, the BER diminishes as the QF increments [25]. BER determines the proportion of bit error to the aggregate number of transmitted bits. In this way, a lower BER demonstrates a better achievement. BER can be influenced by attenuation, noise, scattering, ISI, nonlinear phenomena or even by bit synchronization issues. Furthermore, its performance might be enhanced by propelling a strong signal into a transmission system unless this causes ISI, ICI and more mistakes; by picking a sturdy modulation format, or by applying channel coding plans, among others.

3.2 Eye Diagram

The eye chart is considered a graphical demonstration of signals, in which numerous cycles of the signal are superimposed over each other. The measure of commotion and ISI, which belong to an optical signal, can be estimated from its appearance. Less noise influences the eye to outline look smoother, since there is less mutilation of a signal. The bigger the span of the eye opening is, the lower the error rate will be [25].

3.3 Error Vector Magnitude for M-QAM Signals

It has been noticed that advanced modulation configurations, for example, M-QAM encode information as in amplitude and phase of the optical bearer. Its complex amplitude vector E is got with a coherent receiver and described by M points (symbols) in an intricate constellation plane accordingly. The actually received signal vector Er deviates by an error vector Eerr from the ideal transmitted vector Et. EVM can be estimated from OSNR for M-QAM as [25].

$$EVM_{\text{rms}} = \left[\frac{1}{OSNR} - \sqrt{\frac{96/\pi}{(M-1)OSNR}} \sum_{i=1}^{\sqrt{M}-1} \gamma_i e^{-a_i} + \sum_{i=1}^{\sqrt{M}-1} \gamma_i \beta_i \text{erfc}(\sqrt{\alpha_i}) \right]^{1/2} \tag{14}$$

$$\alpha_i = \frac{3\beta_i^2 OSNR}{2(M-1)} \tag{15}$$

$$\beta_i = 2i - 1, \quad \gamma_i = 1 - \frac{i}{\sqrt{M}} \tag{16}$$

The BER can be approximate as:

$$BER = \frac{1 - M^{-1/2}}{\frac{1}{2}\log_2 M} \text{erfc} \left[\frac{3/2}{(M-1)EVM_{\text{rms}}^2} \right]^{\frac{1}{2}} \tag{17}$$

Equation (13) had been modified in [26] where EVM is used instead of EVMrms. Also the signal to noise ratio (SNR) can be defined from EVM:

$$SNR = \frac{1}{EVM^2} \tag{18}$$

4 Results of Simulation

In this section, optical transmission link for a single channel CO-OFDM system is considered by using co-simulation of VPI software and Matlab. The system simulation settings take into account the most important optical communication system component parameters which include fiber nonlinearity, noise and dispersion, in addition to attenuation.

4.1 CO-OFDM System Settings

For performance investigation, the coherent optical transmission is simulated by the way of using co-simulation of VPI software Ver.9.5 and Matlab 2014a. Likewise, the system simulation settings take into account the most key optical communication system component parameters including fiber nonlinearity, noise, dispersion and attenuation. The scheme in Fig. 2 is used for the purpose of simulating the single channel CO-OFDM system. The obtained OFDM signal is introduced from Matlab and it has been connected with VPI Optical system by using Matlab co-simulation in order to carry data through the optical channel. The system parameters used in the simulation are summarized as follows in Table 1.

Table 1. CO-OFDM system settings parameters

OFDM parameters	
Modulation index	4QAM, 16QAM
Number of subcarrier	128
Oversampling	4
Cyclic prefix	1/8
Bit rate	10 Gbps
Sample rate	40 Gbps
Channel parameters	
SMF length	50 km
DCF length	5 km
Attenuation (SMF)	0.2 dB/km
Attenuation (DCF)	0.5 dB/km
Dispersion (SMF)	16 ps/nm/km
Dispersion (DCF)	−160 ps/nm/km
No. of span	1
Laser parameters	
Carrier frequency	193.1 GHz
Line width for laser	100 kHz
Laser power	2 dBm
Wavelength	1550 nm

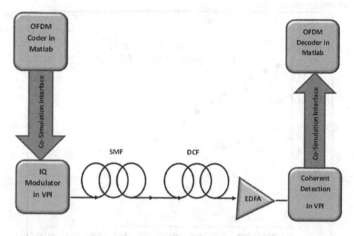

Fig. 2. Simulation scheme of single channel CO-OFDM system

4.2 Performance Evaluation

It has been agreed upon the fact that the system is analyzed and simulated using VPI Software Ver. 9.5 with MATLAB ver. (R2014a) for simulation. The setup has been chosen to build a 10 Gbps transmitted signal. The length of fiber equals to number of loop multiplies by length of (SMF+DCF). Firstly, to investigate the effect of modulation mapping with fixed line width of the CW-laser is chosen 0.1 MHz for certain values of subcarrier and cyclic prefix. Noting that the change in the laser line width generates inherent LPN which imposes a limitation on CO-OFDM system. Thus OFDM system with extended symbol period needs a smaller line width laser [27]. The phase noise along with a line width of the laser will cause a loss of orthogonally which definitely leads to ICI. For this reason, CO-OFDM system uses shorter OFDM symbol than DDO-OFDM system. In this system, we consider Nsc = 128 and CP = 1/8.

Generally speaking, to examine the effect of modulation mapping on the system performance when changing distance, these parameters are considered as follows: N_{sc} = 128, CP = 0.125, and laser line width 0.1 MHz with optimum launch power. Figure 3 demonstrates the relationship between QF and distance in km. Moreover, the parameters for the system as mentioned before but the OSNR is set to 6 dB for 4QAM and later it is set to 15 dB for 16QAM. As we can see from the figure that QF for short distance is good but when the distance increases the QF decreases. Additionally, the QF for 4QAM and 16QAM at a distance 55 km is equal to 16 dB and 11 dB respectively; but at 880 km the QF is equal to 4.8 dB and 0.4 dB respectively. Thus it is found that when we need to increase the value of QF for 16QAM the OSNR must be increased. Figure 4 illustrates the received signal constellation in 4QAM and 16QAM systems for different values of QF and distances. However, it is shown that a good constellation is obtained in the short distances with high and good QF, while the constellation became worse in the long distances and within low QF.

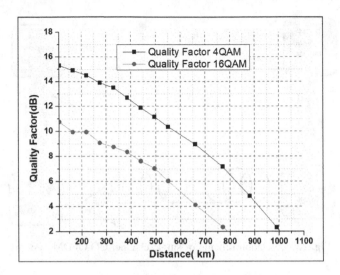

Fig. 3. Quality Factor vs. Distance

Similarly, Fig. 5 refers to the relationship between BER and Distance. The results indicate that when the distance increases, BER increases. It has been illustrated that the BER performance for 4QAM can achieve values below 10^{-3} for a distance between 55 km to 400 km but for 16QAM it becomes below 10^{-1}. On the other hand, the BER for M-QAM is proportional to the OSNR as mentioned in [20]. Thus, for 4QAM if BER is demanded below 10^{-3} the OSNR should be 13 dB, while in 16QAM the OSNR can be more than 15 dB to obtain BER below 10^{-3}. This result indicates that for higher order of M-QAM, higher OSNR is demanded to get a lowest BER below 10^{-3}. In consequence, it can be concluded that the obtained results of the foregoing systems are

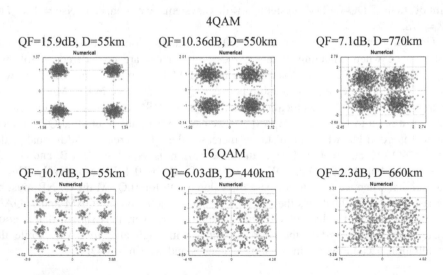

Fig. 4. Signal constellation in 4QAM and 16QAM for different QF and distance

different from the findings of the existing systems because of the use of DCF to compensate the CD of the channel instead of using channel estimation as in [20]. Also the current results are different from the finding of the existing system in [18, 19] because of the use of different value for system setup like power of laser. Whatever the case, when the power of laser increases, the CD in optical channel increases too that leads to higher BER.

Figure 6 illustrates the relation between EVM and the distance in km for 4QAM and 16QAM, respectively. It is clearly noticed that EVM nearly becomes fixed up to 850 km for 4QAM and 550 km for 16QAM, respectively. Meanwhile, when the distance increases the EVM drastically increases causing high BER at the end receiver.

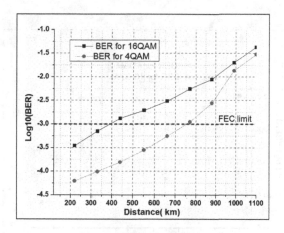

Fig. 5. BER vs. Distance

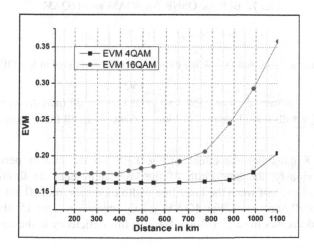

Fig. 6. EVM vs. Distance

Then again, the current paper has focused on the received OSNR in the case of changing in the laser line width. Thus, Fig. 7 shows OSNR vs. BER for the modulation mapping of 4QAM and 16QAM, respectively when the line width is 0.1 MHz. The obtained results reveal that whenever OSNR increases the corresponding BER become smaller. As a result, Table 2 summarizes the comparison performance between 4QAM and 16QAM for the required BER floor and the corresponding OSNR. In the 4QAM, when the OSNR is 8.5 dB, the BER of the system is 10^{-3}. In 16QAM, the OSNR increases up to 7 dB but with 10^{-1} BER. In contrast, when BER becomes 10^{-2} then the required OSNR should increase more. In the case of 16QAM to achieve BER 10^{-3}, the value of OSNR should be greater than 18 dB compared with 8.5 dB for 4QAM scheme.

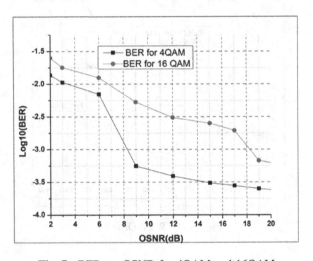

Fig. 7. BER vs. OSNR for 4QAM and 16QAM

Table 2. Minimum received OSNR for two modulation mappings in CO-OFDM system

BER	4QAM	16QAM
10^{-2}	OSNR $\geq$ 3 dB (max distance = 990 km)	OSNR $\geq$ 7 dB (max distance = 660 km)
10^{-3}	OSNR $\geq$ 8.5 dB (max distance = 770 km)	OSNR $\geq$ 18 dB (max distance = 275 km)

Now, Fig. 8 illustrates the effect of the LPN on the system performance. As mentioned previously, the line width of the laser represents the limitation in CO-OFDM system. Note that when the laser line width increases from 10 kHz to 2.2 MHz, the QF dramatically decreases. For 4QAM the QF decreases from 15 dB to 4 dB and for 16QAM it decreases from 11 dB to −0.6 dB; this deficiency in the result arises due

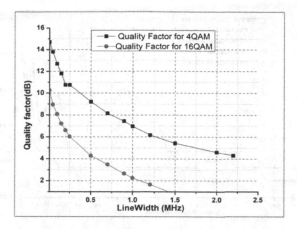

Fig. 8. Quality Factor vs. Laser Line Width

to the fiber nonlinearity and dispersion. Therefore, a good QF can be obtained when the laser line width is below 0.2 MHz for 16QAM. For 4QAM at line width equal to 0.72 MHz, QF achieves 8.

On the other hand, to investigate the effect of laser line width in 4QAM and 16QAM CO-OFDM system which is classified with higher bandwidth efficiency compared to 4QAM. Figure 9(a, b) shows the relationship between EVM and input optical power which is changed from −10 dBm to 20 dBm for mapping order 4QAM and 16QAM at a distance 50 km. It has been clearly noticed that the required optimum laser power for 4 and 16QAM system dramatically varies in the nonlinear oscillated form to achieve low EVM when the laser line width of the laser changes. Thus Table 3 can summarize some of the required optimum laser power for different laser line widths that the system will operate in a good performance.

Nonetheless, Fig. 10(a, b) depicts the relationship between the QF and the distance for the two 4QAM and 16QAM, respectively, different laser line width and the required optimum launch power in dBm. The required optimum launch laser power is also obtained in Table 3 for various ranges of line width to guarantee the successful transmission. The results clearly show that the laser line width has a significant impact on the QF of the system in 4QAM and 16QAM mapping. The higher laser line width means the lower quality performance. The quality penalty is nearly 3 dB to 3.5 dB for 16QAM compared to 4QAM over a maximum allowable distance greater than 850 km–1000 km. On average, it is found that 0.1 MHz–0.4 MHz laser line width needs optimum power −3 dBm to −4 dBm in 16QAM; meanwhile 1.2 MHz and 2 MHz can faithfully operate at −8 dBm compared to 4QAM at 14 dBm or 3 dBm. Thus we can conclude that the best QF at the lowest BER and greater distance can be obtained when the laser line width is chosen 0.1 MHz in both 4QAM operating at −1 dBm and 16QAM at −4 dBm as shown in Fig. 10.

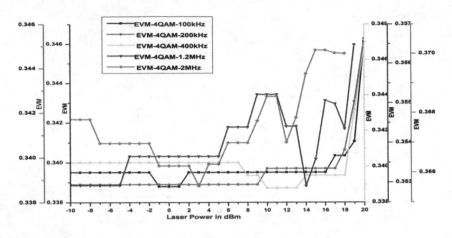

(a)

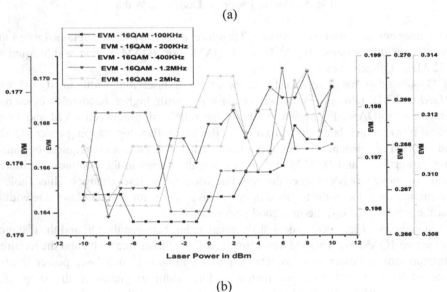

(b)

Fig. 9. EVM vs. Laser Power in dBm (a) 4QAM (b) 16QAM

Table 3. Optimum laser power ranges in 4QAM and 16QAM for different laser line widths

Line width of laser	Optimum power in dBm	
	16QAM	4QAM
0.1 MHz	−6, −5, −4, −3, −2, −1	−1, 0, 1
0.2 MHz	−3, −2, −1, 0	−10 to 9
0.4 MHz	−8, −7, −6, −5, −4	10, 11, 12, 13
1.2 MHz	−8	14
2 MHz	−9, −8	3

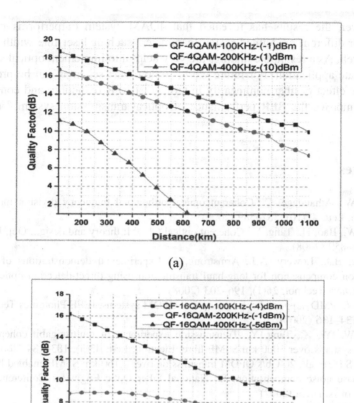

(a)

(b)

Fig. 10. QF vs. Distance (a) 4QAM (b 16QAM

5   Conclusions and Discussion

This article has evaluated the performance of M-QAM modulated CO-OFDM system including the effect of laser line width in addition to changing the power of laser. The system is significantly designed by using OFDM signals of 10 GB/s data rate transfer over standard SMF. Meanwhile the effect of chromatic-dispersion is compensated by using DCF compensating technique. The result clearly indicates that performance of CO-OFDM system is affected by the change of laser line width, the power of laser and the order of QAM mapping. Thus the system design has been checked through the metrics of QF, BER, EVM and OSNR, considering that the choice of laser line width should be narrower to get optimum system performance. The EVM analysis shows that the input optical power of the laser from −6 dBm to 6 dBm can clearly provide a robust and good performance for both systems of 4QAM and 16QAM.

Moreover, the results has revealed that 4QAM system outperforms more than 16QAM for different predefined system parameters such as laser line width and laser power as well. As a result, the findings are very helpful to implement optical system for high data rate application. For future work, some robust schemes can be proposed to mitigate the effect of fiber nonlinearity in terms of PAPR reduction; and consequently they may improve the BER performance of optical transmission system for high bit rate.

References

1. Shieh, W., Athaudage, C.: Coherent optical orthogonal frequency division multiplexing. Electron. Lett. **42**(10), 587–589 (2006)
2. Shieh, W., Bao, H., Tang, Y.: Coherent optical OFDM: theory and design. Opt. Express **16** (2), 841–859 (2008)
3. Schmidt, B.J., Lowery, A.J., Armstrong, J.: Experimental demonstrations of electronic dispersion compensation for long-haul transmission using direct-detection optical OFDM. J. Lightwave Technol. **26**(1), 196–203 (2008)
4. Shieh, W.: PMD-supported coherent optical OFDM systems. IEEE Photonics Technol. Lett. **19**(3), 134–136 (2007)
5. Shieh, W., Yi, X., Tang, Y.: Transmission experiment of multi-gigabit coherent optical OFDM systems over 1000 km SSMF fibre. Electron. Lett. **43**(3), 183–184 (2007)
6. Jansen, S.L., et al.: 20-Gb/s OFDM transmission over 4,160-km SSMF enabled by RF-pilot tone phase noise compensation. In: National Fiber Optic Engineers Conference. Optical Society of America (2007)
7. Yi, X., Shieh, W., Ma, Y.: Phase noise effects on high spectral efficiency coherent optical OFDM transmission. J. Lightwave Technol. **26**(10), 1309–1316 (2008)
8. Peng, W.-R.: Analysis of laser phase noise effect in direct-detection optical OFDM transmission. J. Lightwave Technol. **28**(17), 2526–2536 (2010)
9. Nazarathy, M., et al.: Recent advances in coherent optical OFDM high-speed transmission. In: PhotonicsGlobal@ Singapore. IPGC 2008. IEEE (2008)
10. Shoreh, M.H.: Compensation of nonlinearity impairments in coherent optical OFDM systems using multiple optical phase conjugate modules. J. Opt. Commun. Netw. **6**(6), 549–558 (2014)
11. Ip, E., Kahn, J.M.: Digital equalization of chromatic dispersion and polarization mode dispersion. J. Lightwave Technol. **25**(8), 2033–2043 (2007)
12. Kudo, R., et al.: Coherent optical single carrier transmission using overlap frequency domain equalization for long-haul optical systems. J. Lightwave Technol. **27**(16), 3721–3728 (2009)
13. Shieh, W., Ho, K.-P.: Equalization-enhanced phase noise for coherent-detection systems using electronic digital signal processing. Opt. Express **16**(20), 15718–15727 (2008)
14. Goebel, B., et al.: PAPR reduction techniques for coherent optical OFDM transmission. In: 11th International Conference on Transparent Optical Networks. ICTON 2009. IEEE (2009)
15. Dung, H.V.T., et al.: PAPR reduction using PTS with low computational complexity in coherent optical OFDM systems. In: 2012 18th Asia-Pacific Conference on Communications (APCC), IEEE (2012)
16. Rishi, P., Tamilselvi, S.: Mitigation of non linear effects using non linear transform in dispersion managed coherent optical OFDM systems. J. Comput. Theor. Nanosci. **15**(2), 551–557 (2018)

17. Wang, H., et al.: Performance evaluation of (D) APSK modulated coherent optical OFDM system. Opt. Fiber Technol. **19**(3), 242–249 (2013)
18. Saini, S.S., Sheetal, A., Singh, H.: Coherent optical OFDM system or Long-Haul transmission. IJIET (2015)
19. John, M.A.J., Gokul, P.: Performance evaluation and simulation of OFDM in optical communication systems. Int. J. Eng. Res. Appl. **5**(2), 01–04 (2015)
20. Wang, Z., et al.: Performance analysis of different modulation schemes for coherent optical OFDM system. In: International Conference on Communications and Mobile Computing (CMC) 2010. IEEE (2010)
21. Nehra, M., Kedia, D.: Design of optical I/Q modulator using dual-drive Mach-Zehnder modulators in coherent optical-OFDM system. J. Opt. Commun. **39**(2), 155–159 (2018)
22. Gnanagurunathan, G., Rahman, F.A.: Comparing FBG and DCF as dispersion in the long haul narrowband WDM systems. In: 2006 IFIP International Conference on Wireless and Optical Communications Networks. IEEE (2006)
23. Norgard, G.: Chromatic dispersion compensation: extending your reach. http://www.datacenterjournal.com/chromatic-dispersion-compensation-extending-reach/. 18 April 2016
24. Tang, J., Lane, P., Shore, K.A.: High-speed transmission of adaptively modulated optical OFDM signals over multimode fibers using directly modulated DFBs. J. Lightwave Technol. **24**(1), 429 (2006)
25. Freude, W., et al.: Quality metrics for optical signals: eye diagram, Q-factor, OSNR, EVM and BER. In: 2012 14th International Conference on Transparent Optical Networks (ICTON). IEEE (2012)
26. Sanya, M.F., et al.: Performance analysis of known unipolar optical OFDM techniques in PON IM/DD fiber link. In: 2013 10th International Conference on High Capacity Optical Networks and Enabling Technologies (HONET-CNS). IEEE (2013)
27. Pawar, V., Umbardand, A.U.: Performance evaluation of high speed coherent optical OFDM system. Int. J. Current Trends Eng. Res. (IJCTER) **2**(7), 129–142 (2016)

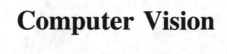

Computer Vision

Enhanced Ensemble Technique for Optical Character Recognition

Imad Qasim Habeeb[1]([⊠]) [iD], Zeyad Qasim Al-Zaydi[2] [iD],
and Hanan Najm Abdulkhudhur[3] [iD]

[1] Engineering College,
University of Information Technology and Communications, Baghdad, Iraq
emadkassam@uoitc.edu.iq
[2] Biomedical Engineering, University of Technology, Baghdad, Iraq
11344@uotechnology.edu.iq
[3] Ministry of Higher Education and Scientific Research (MOHESR),
Baghdad, Iraq
hanan_nagem@yahoo.com

Abstract. Optical character recognition (OCR) is the electronic transformation of images into a computer-encoded text. OCR systems often produce poor accuracy for noisy images. Ensemble recognition techniques are used to improve OCR accuracy. The idea of the ensemble recognition techniques is to produce N-versions of an input image. These versions are similar but not identical. They are passed through the OCR engine to turn them into different OCR outputs, which later leads to select the best between them. Existing ensemble techniques need to be more effective to reduce OCR error rate. This research proposed enhanced ensemble technique to overcome the drawbacks of existing techniques. The proposed technique was evaluated against three other relevant existing techniques. The performance measurements used in this research were Word Error Rate (WER) and Character Error Rate (CER). Experimental results showed a relative decrease of 14.37% and 40.13% over the WER and CER of the best existing technique. This study contributes to the OCR domain as the proposed technique could facilitate the automatic recognition of documents. Hence, it will lead to a better information extraction.

Keywords: Ensemble technique · Noisy images
Automatic document recognition · OCR

1 Introduction

The accuracy of OCR systems is still considered an open problem in the following cases. The first case is for the cursive written-based languages [1, 2]. The second case is for low scanning resolution image [3–5]. The last case is when the image contains noise [6, 7]. It is difficult to identify OCR error rate accurately for previous cases. The reason is that the error rate of OCR changes depends on type and size of testing dataset [8], scanning resolution of an image [5, 9], and lastly, noise types in the image [10]. For example, the error rate of OCR for noise images can even reach up to 100% [11].

© Springer Nature Switzerland AG 2018
S. O. Al-mamory et al. (Eds.): NTICT 2018, CCIS 938, pp. 213–225, 2018.
https://doi.org/10.1007/978-3-030-01653-1_13

According to Al Azawi [12] and Volk, Furrer [13], using ensemble techniques to produce multiple outputs of OCR is better than using single OCR output in reducing the error rate. This fact has been supported by many researchers [5, 12, 14].

The idea of ensemble techniques is to look for differences between several outputs of OCR and then choose the best among them. This approach incorporates three processes: ensemble, alignment, and voting [14]. Figure 1 shows an example of OCR multiple outputs approach.

Reference Text =	"Google	chrome	is	a	free"
Output of OCR 1=	"Googie	chroome	is	a	free"
Output of OCR 2=	"Gcogle	chronne	1		fne"
Output of OCR 3=	"Google	chronne	ls	a	fiie"

Alignment of three different OCR outputs=

G	o	o	g	i	e		c	h	r	o	o	m	e		i	s		a		f	r	e	e
G	c	o	g	l	e		c	h	r	o	n	n	e		1	-		-		f	n	-	e
G	o	o	g	l	e		c	h	r	o	n	n	e		1	s		a		f	i	i	e

Fig. 1. OCR multiple outputs example

From this example, it can be noted that ensemble process was used to produce 3-versions of the input image. These versions entered the OCR engines to produce 3-different outputs. However, Fig. 1 also shows that a different number of characters were resulting from three distinct OCR engines. This causes vertical words overlapping between the OCR resulting texts. Therefore, an alignment process is required to parallel each character with corresponding in other OCR outputs. After the alignment process, a voting process is used to select the best word between the multiple outputs of the OCR.

Results from earlier studies [2, 6, 14], showed that the multiple thresholds technique is one of the best choices among them. Figure 2 shows an example of the multiple thresholds technique proposed by [13]. In this technique, [13] produced seven images using seven threshold values to the binarization for a single image. The seven images were passed to the same OCR engine to turn them into seven OCR outputs. The threshold values are 31, 63, 95, 127, 159, 191, and 223. From Fig. 2, it can be seen that some important features of the characters' images are lost especially for threshold values of 31, 63, 191, and 223. This is because the pixels values above the identified threshold are transformed as white and the ones that are larger than the threshold become black. For example, when using threshold value 31, the image will become close to the white. For threshold value 63, the image will become more intense. Likewise, for threshold value 191, the image will become more intense. For threshold value 223, the image will become close to the black. Thus, these threshold values result in losing some important features of the characters' images. This causes an increasing number of wrong words in the OCR outputs. Therefore, this research enhanced Multiple Thresholds technique so that the number of wrong words in the OCR outputs can be reduced. The contribution to the OCR domain could facilitate the automatic recognition of noisy documents.

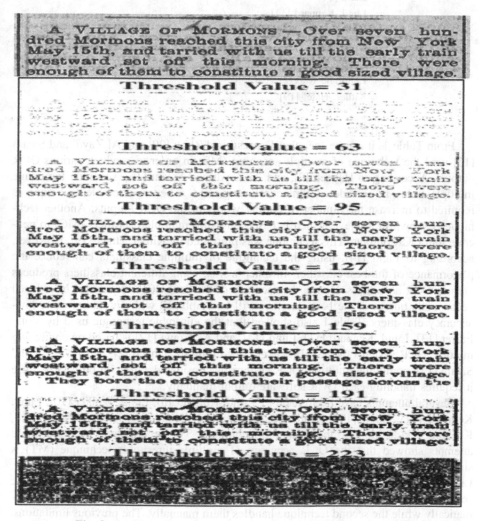

Fig. 2. Multiple thresholds technique example. Source: Lund [14]

2 Related Work

Table 1 shows existing ensemble techniques used during the ensemble process that belong to the multiple outputs of OCR.

Table 1. Existing ensemble techniques used in multiple outputs of OCR.

Author	MT	MOS	MC	MS
Al Azawi [12]		✓		
Al-Zaydi and Salam [15]				✓
Lund [14]	✓			
Lund, Kennard [11]	✓			
Batawi and Abulnaja [16]		✓		
Volk, Furrer [13]		✓		
Lund and Ringger [17]		✓		
Lund and Ringger [18]		✓		
Kittler, Hatef [19]			✓	

From Table 1, it can be seen that Lopresti and Zhou [20] and Al-Zaydi and Salam [15] used the technique of multiple scanning (MS) in order to produce various OCR outputs. According to Habeeb [2], and Al-Zaydi and Salam [15], scanning the image multiple times is considered time-consuming, and a better ensemble technique is required to increase the number of correct words in the OCR outputs. Another technique is shown in Table 1, which is based on using multiple classifiers (MC) to generate four OCR outputs. This technique was presented by Kittler, Hatef [19]. However, according to Lund [14], using the technique of multiple classifiers can reduce the performance of the best classifier. In other words, using different classifiers produces various OCR accuracies. Hence, the accuracy of the final OCR output may be reduced because the output of the best classifier will be combined with outputs of other low-accuracy classifiers. Table 1 also showed that the ensemble technique used by Volk, Furrer [13], Lund, Kennard [11], Lund [14], and Al Azawi [12] is based on combining multiple OCR systems (MOS). However, According to Habeeb [2], and Al-Zaydi and Salam [15], combining multiple OCR systems is considered a difficult process because it requires handling each OCR software output manually.

Several attempts have been made by Lund, Ringger [6], Lund, Walker [10], Lund, Kennard [11], Lund [14], Lund and Ringger [17], Lund and Ringger [18], Lund, Kennard [21] to improve the ensemble techniques. The experimental results of these attempts showed that the OCR accuracy of multiple thresholds technique (MT) is similar to the OCR accuracy obtained by combining different OCR software. However, Lund [14] proved that using the multiple thresholds technique is easier than using the second technique. This is because the first technique handles OCR outputs programmatically while the second technique handles them manually. The previous limitations, which are mentioned above, explained that an enhanced technique is needed to reduce these limitations.

This research adopted the multiple thresholds technique as proposed by Lund [14] to be enhanced. The reasons are: firstly, it does not require scanning image several times. Secondly, it does not require combining different classifiers that can reduce the performance of the best classifier as mentioned by Lund [14]. Lastly, it does not require combining multiple OCR systems that are considered as a manual process by Al-Zaydi and Salam [15]. The next section explains the steps undertaken to enhance this technique.

3 Proposed Technique

The proposed technique has been referred to as EET by this research, which means enhanced ensemble technique. As mentioned previously, the EET enhanced Multiple Thresholds technique of Lund [14]. The enhancement was performed depending on six steps. (Step 1) develop an algorithm for Multiple Thresholds technique based on the description of this technique in [14]. Equation 1 shows the mathematical expression of Multiple Thresholds technique [14]:

$$PV[i, j] = \begin{cases} 0 & \text{if } PV[i, j] \leq x \\ 255 & \text{if } PV[i, j] > x \end{cases} \tag{1}$$

Where i and j are the counters of an image array, PV[i, j] is a pixel's value at [i, j], and x is a threshold value used to convert an image to black and white. The threshold values of x are 31, 63, 95, 127, 159, 191, and 223 so that multiple images were produced and passed to the OCR engines to turn them into seven OCR outputs. For example, the threshold value of x = 31 was used to generate version-1 of the input image while threshold value of x = 63 was used to generate version-2 of the input image, and so on. (Step 2) run the algorithm of Multiple Thresholds technique using a dataset of noisy images for each threshold value (31, 63, 95, 127, 159, 191, and 223) separately.

(Step 3) for each threshold value mentioned in Step 3, the OCR error rate was measured. (Step 4) based on the results of the OCR error rate, this research selected the best three threshold values, which are (127, 159, and 191), to be used in the proposed technique. (Step 5) round the threshold values of 127, 159, and 191 to 130, 160, and 190 respectively, to make the difference between them the same. Hence, the proposed ensemble technique generated only 3-versions from the original image. For example, the threshold value of x = 130 was used to generate version-1 of the input image while threshold value of x = 160 was used to generate version-2 of the input image, and so on. The generated versions are similar, but not identical. (Step 6) it was found that Multiple Thresholds technique has a problem, which is a loss of important features from characters images because this technique changed some grey pixels values (above threshold) to white (255). (Step 7) based on the problem in Step 6, an improvement was made to handle this problem. The improvement modified Eq. 1 mentioned previously to be as Eq. 2:

$$PV[i, j] = \begin{cases} 0 & \text{if } PV[i, j] \leq x \\ PV^{\wedge} & \text{if } x < PV[i, j] < (x+30) \\ 255 & \text{if } PV[i, j] \geq 220 \end{cases} \tag{2}$$

where i and j are the counters of an image array, $PV[i, j]$ is a pixel's value at $[i, j]$, $PV^{\wedge}$ is a pixel's value after performing an ensemble cycle (described later) and x is a threshold value (130, 160, and 190) used to convert an image to black and white. From Eq. 2, it can be seen that the proposed technique divided pixels' values of an image into three groups. The first group contains the pixels that have values close to the black, which range from (0 to x). The second group contains the pixels that have values close to the white, which range from (220 to 255). The last group contains weak values that lie between the first and second group, which range between (x + 1 and x + 29).

The goal of dividing pixels' values into groups is to confirm the first group by making its members black and confirm the second group by making its members white while the last group will be subjected to an ensemble cycle. The purpose of the ensemble cycle is to modify grey pixels values along the path of black pixels by performing several operations called a cycle on them. In other words, the proposed technique focuses on grey pixels in the path of black pixels because white pixels do not represent any information about images of characters while black pixels give strong evidence that they may represent information about them. Therefore, if the ensemble cycle was implemented in different ways, then this can lead toproducing differences between resulting versions of the input image.

To clarify more, a variable named x is used as an initial step to achieve the proposed technique goal of dividing pixels' values of an image into three groups. The value of x is changed during the production of each image. It takes the values 130, 160, and 190 to produce image 1, 2, and 3 respectively. The values of x are selected based on two factors. The first factor is that they should lie between black's pixels and white's pixels. Therefore, the proposed technique selected values of x between 130 and 220. This is because pixels' values under 130 are close to the black, and pixels' values greater than 220 are close to the white. The second factor is that this research conducted a series of experiments as described earlier, and based on results, it found that the best values of x are 130, 160, and 190. To produce any version of the original image, several operations are performed, called an ensemble cycle, as shown in Fig. 3.

Firstly, each pixel's value that is equal to or smaller than x will change to zero, while each pixel's value that is equal to or greater than 220 will change to 250. The reason for this is to confirm the stronger pixels. Secondly, any pixel's value that is greater than x and less than 220 will remain unchanged until performing ensemble cycle on them. The ensemble cycle will start by identifying all pixels having values between (x + 1) and (x + 29), located beside pixels having values equal to zero as shown in part A in Fig. 3. These will become primary starting pixels for the ensemble process. In the proposed ensemble cycle, each primary starting pixel has a cycle of operations: (1) value of starting pixel is changed to zero, and (2) all the neighbouring pixels from all sides with non-zero values are arranged in ascending order, so that the

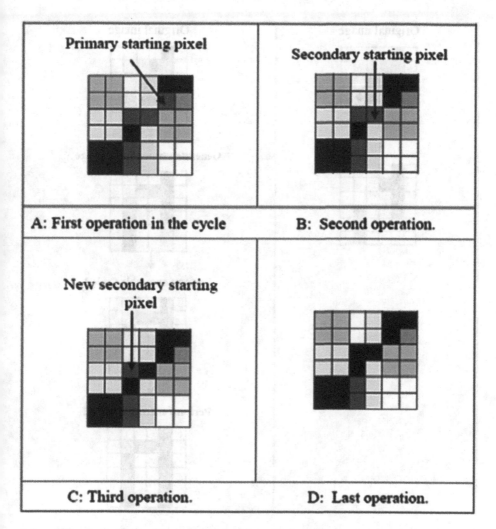

Fig. 3. A simple example on ensemble cycle for one primary starting pixel.

pixel having the smallest value becomes a secondary starting pixel, on condition that its value lies between $(x + 1)$ and $(x + 29)$. If these conditions are met, then all previous operations (1 and 2) are performed for the new starting pixel and so on, otherwise the cycle is ended for the current starting pixel, and another cycle for the next primary starting pixel is initiated. At the end of the ensemble cycle of each primary starting pixel, the value of a single pixel or the values of multiple pixels are changed based on the proposed ensemble cycle. Figure 4 shows why the proposed ensemble technique is better than existing technique proposed by Lund [14].

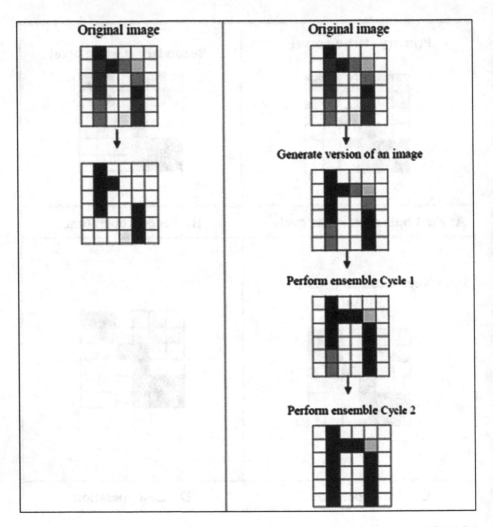

Fig. 4. An image resulted by using the proposed technique with a threshold value = 160 (right), and resulting image by using the existing technique with threshold value = 160 (left).

From Fig. 4, it can be seen that the resulting image produced by the existing technique proposed by Lund [14] is not similar to the original image while the resulting image produced by the proposed technique is similar. As mentioned previously, this is because the existing technique makes all pixel values above threshold value white. This will lead to loss of some important features from characters' images [15]. The effect of losing some features from characters' images is that the number of wrong words in OCR outputs will be increased. In contrast, the proposed technique preserves features of characters' images. Furthermore, it restores them to the original shape. Algorithm 1 represents the pseudo code for proposed technique.

Algorithm 1: Proposed technique	
S1	Let z is an array of [130, 160, 190]
S2	Let k=0 // counter for elements in the array z
S3	Let x =z [k] // x is used to save the threshold value
	// Produce N-versions of dataset images
S4	Let c=0 // counter for images in the dataset images
S5	Select image[c] from the dataset images
S6	If each pixel value <= x, then this value will be changed to zero
S7	Let PSP is an array of pixels having values between (x+1) and (x+29), located beside pixels having values equal to zero
S8	Let y= 0 // counter for elements in the array PSP
S9	Let Pixel(i, j) refers to the position of PSP[y] in image[c]
S10	Pixel(i, j)=0
S11	Let SSP is an arrayof neighbouring pixels to the Pixel(i, j) from all sides with non-zero values
S12	Let h= minimum value in array SSP

S13	If the value of h between (x+1) and (x+29) then
	    Let Pixel(i, j) refers to the position of minimum value in array SSP in the image[c] and go to S10
	else if y is the last element in PSP array then
	go to S14
	else
	    y=y+1 and go to S9

S14	Save image[c] with different name // to maintain original dataset images

S15	if the image[c] is the last one in the dataset images then
	go to S16
	else
	c=c+1 and go to S5

S16	if z [k] is the last element in the array of z then
	go to S17
	else
	k=k+1 and go to S3

S17	End

From step S1 of Algorithm 1, it can be seen that the proposed technique used an array called "z", which has three elements 130, 160, and 190. These elements were used to produce 3-versions of testing dataset images as explained previously. Steps S2 to S4 define the variables used in this algorithm. In Step S5, each image from the testing dataset was selected in order to process by the proposed technique. In step S6, if the value of each pixel $\leq$ x, then this value will be changed to zero. After that, some grey pixels values will be changed to the black if they located in the path of black pixels (Steps S7 to S14). As mentioned previously, the proposed technique focuses on grey pixels in the path of black pixels because white pixels do not represent any information about images of characters while black pixels give strong evidence that they may represent information about them. This study finds the path of black pixels by focusing only on grey pixels that must be located beside black pixels while other grey pixels that do not satisfy this condition will be ignored. In Step S15, the next image from the dataset images would be selected to be handled by this technique while in step S16 all previous steps are repeated for the next element in the array "z".

After performing Algorithm 1, three similar but non-identical dataset images are produced. These three dataset images will be sent to the three versions of the same OCR engine in order to turn them into three outputs text. The voting process will select the best between them to produce a single OCR output text.

4   Experimental Results

In this section, the proposed ensemble technique EET has been implemented. Furthermore, three related existing techniques have also been implemented to be used in the evaluation of EET. They are used by Al Azawi [12], Lund [14], and Al-Zaydi and Salam [15]. In addition to that, two metrics have been used during the evaluation process. They are word error rate (WER) and character error rate (CER). Equations 3 and 4 show how to compute the WER and CER respectively [2, 22].

$$\text{WER} = \frac{\text{Wrong words}}{\text{Total words}} * 100 \tag{3}$$

$$\text{CER} = \frac{\text{Wrong characters}}{\text{Total characters}} * 100 \tag{4}$$

Wrong words and characters of Eqs. 3 and 4 have been counted using Levenshtein distance [22]. The testing dataset used during the evaluation has five characteristics. It contained 50710 characters and symbols in the form of 70 noisy documents. It was chosen randomly from old books. It contained, in addition to the characters of a text, the special symbols, such as commas and brackets. It included eight different fonts and for each font, six different sizes ranging from 10 to 20 were included. The texts in these documents acted as a reference text during the evaluation process. The hardcopy was scanned at 300 dpi with a grey level in a modern scanner to produce the testing dataset images. The experimental results of the EET evaluation using the WER metric are shown in Table 2.

From Table 2 it can be clearly seen that the WER value for each ensemble technique is different from the others. Overall, they show that the technique of Lund [14] had the highest percentage value of WER with the rate of 42.34%. This is followed by the technique of Al Azawi [12] 38.79%, and technique of Al-Zaydi and Salam [15] 36.02%. Furthermore, it can be seen that WER of the proposed technique EET had the lowest percentage value of OCR error rate than the others with the rate of 30.84%. This technique has a 20.67% relative decrease on the mean WER of the three existing ensemble techniques and 14.37% relative decrease on the best WER of them. The relative decrease in WER has been measured using Eq. 3. This indicates that the proposed technique EET had a reduction in the WER metric compared to the existing techniques.

Table 2. Experimental results of the EET evaluation using the WER metric.

	Al-Zaydi and Salam [15]	Lund [14]	Al Azawi [12]	EET
Total words	10142	10142	10142	10142
Wrong words	3653	4294	3934	3128
WER	36.02%	42.34%	38.79%	30.84%

On the other hand, the experimental results of the EET evaluation using the CER metric are shown in Table 3.

Table 3. Experimental results of the EET evaluation using the CER metric.

	Al-Zaydi and Salam [15]	Lund [14]	Al Azawi [12]	EET
Total characters	50710	50710	50710	50710
Wrong characters	6324	8543	6483	3786
CER	12.47%	16.85%	12.78%	7.47%

As can be seen in Table 3, the ensemble techniques of Al-Zaydi and Salam [15] and Al Azawi [12] show a slight difference in the values of CER with rates of 12.47% and 12.78% respectively. Furthermore, the CER values for both previous techniques are less than CER value for the technique of Lund [14], which has a rate of 16.85%. The proposed ensemble technique EET outperformed the existing techniques in terms of CER with the rate of 7.47%. The proposed technique has a 45.81% relative decrease on the mean CER of the three existing ensemble techniques and 40.13% relative decrease on the best CER of them. The previous values of CER show that the proposed technique EET had the highest percentage decrease in the number of wrong characters compared to the examined existing techniques.

There are three advantages resulting from the proposed ensemble technique. The first advantage is that it does not require scanning an image multiple times which is considered a boring process as mentioned by Al-Zaydi and Salam [15]. The second advantage is that it does not require connecting different OCR systems, considered a difficult and manual process as stated by Al-Zaydi and Salam [15]. The last advantage is that it is better than the existing technique proposed by Lund [14] that generates seven outputs by using seven threshold values as mentioned previously.

5 Conclusion

The idea of ensemble techniques when used in OCR applications is to produce differences between several outputs of OCR and then choose the best among them. Therefore, the goal of this research is to improve OCR accuracy for the noisy images by proposing solutions to the limitations of existing ensemble techniques. Hence, this study has designed an EET technique, which is based on the concept of changing some grey pixels values that located only in the path of black pixels if they met certain conditions. The proposed technique focuses on grey pixels in the path of black pixels because black pixels give strong evidence that they may represent information about characters. In contrast, existing techniques focus on modifying all pixels values in an image, which may cause noises if some grey pixels values converted to black, and may cause loss of important information if some grey pixels values converted to white.

Experimental results proved that the concept proposed by this study is able to decrease an OCR error rate for noisy images. The proposed technique EET outperforms other existing related techniques in terms of WER and CER. Therefore, the practical results of this study indicate that the main objective of it is achieved. The experiments also show that the error rate is high for noisy images. This presents a fact that it is difficult for OCR accuracy to be 100% for this type of images. High OCR error rate values of this study are similar to the high error rate values mentioned in previous related studies, such as results of Lund [14]. The proposed EET hopes to contribute various topics that initially employed the ensemble approach, and this includes low-resolution images and cursive written-based languages. Future research of this study is to design techniques that can reduce existing limitations in the alignment and voting processes, which belongs to the ensemble approach of OCR systems.

References

1. Habeeb, I.Q., Yusof, S.A., Ahmad, F.B.: Two bigrams based language model for auto correction of arabic OCR errors. Int. J. Digit. Content Technol. Appl. **8**(1), 72–80 (2014)
2. Habeeb, I.Q.: Hybrid model of post-processing techniques for arabic optical character recognition. Universiti Utara Malaysia, Kedah, Malaysia (2016)
3. Ma, D., Agam, G.: A super resolution framework for low resolution document image OCR. In: Proceedings of the International Society for Optical Engineering (SPIE) on Document Recognition and Retrieval XX. International Society for Optics and Photonics, California, USA (2013)
4. Ma, D., Agam, G.: Lecture video segmentation and indexing. In: Proceedings of the International Society for Optical Engineering (SPIE) on Document Recognition and Retrieval XIX. International Society for Optics and Photonics, California, USA (2012)
5. Habeeb, I.Q., Yusof, S.A., Ahmad, F.B.: Improving optical character recognition process for low resolution images. Int. J. Adv. Comput. Technol. **6**(3), 13–21 (2014)
6. Lund, W.B., Ringger, E.K., Walker, D.D.: How well does multiple OCR error correction generalize? In: Proceedings of Document Recognition and Retrieval XXI (DRR 2014). International Society for Optics and Photonics, San Francisco, USA (2014)
7. Herceg, P., et al.: Optimizing OCR accuracy for bi-tonal, noisy scans of degraded arabic documents. In: Proceedings of the International Society for Optical Engineering (SPIE) on Visual Information Processing. SPIE - The International Society for Optical Engineering, Florida, USA (2005)
8. Bassil, Y., Alwani, M.: OCR post-processing error correction algorithm using google online spelling suggestion. J. Emerg. Trends Comput. Inf. Sci. **3**(1), 90–99 (2012)
9. Al-Zaydi, Z.Q., Vuksanovic, B., Habeeb, I.Q.: Image processing based ambient context-aware people detection and counting. Int. J. Mach. Learn. Comput. (IJMLC) **8**(3), 268–273 (2018)

10. Lund, W.B., Walker, D.D., Ringger, E.K.: Progressive alignment and discriminative error correction for multiple OCR engines. In: Proceedings of the 11th International Conference on Document Analysis and Recognition (ICDAR 2011). IEEE, Beijing (2011)

11. Lund, W.B., Kennard, D.J., Ringger, E.K.: Why multiple document image binarizations improve OCR. In: Proceedings of the Workshop on Historical Document Imaging and Processing (HIP 2013). ACM, Washington (2013)

12. Al Azawi, M.: Statistical Language Modeling for Historical Documents Using Weighted Finite-State Transducers and Long Short-Term Memory. Technical University of Kaiser-slautern, Kaiserslautern (2015)

13. Volk, M., Furrer, L., Sennrich, R.: Strategies for reducing and correcting OCR errors. In: Sporleder, C., van den Bosch, A., Zervanou, K. (eds.) Language Technology for Cultural Heritage. Theory and Applications of Natural Language Processing. Springer, Heidelberg (2011). https://doi.org/10.1007/978-3-642-20227-8_1

14. Lund, W.B.: Ensemble Methods for Historical Machine-Printed Document Recognition. Brigham Young University, Utah (2014)

15. Al-Zaydi, Z.Q., Salam, H.: Multiple outputs techniques evaluation for arabic character recognition. Int. J. Comput. Tech. (IJCT) **2**(5), 1–7 (2015)

16. Batawi, Y., Abulnaja, O.: Accuracy evaluation of arabic optical character recognition voting technique: experimental study. Int. J. Electr. Comput. Sci. **12**(1), 29–33 (2012)

17. Lund, W.B., Ringger, E.K.: Error correction with in-domain training across multiple OCR system outputs. In: Proceedings of the 11th International Conference on Document Analysis and Recognition (ICDAR 2011). IEEE, Beijing (2011)

18. Lund, W.B., Ringger, E.K.: Improving optical character recognition through efficient multiple system alignment. In: Proceedings of the 9th ACM/IEEE-CS Joint Conference on Digital Libraries. ACM, Austin (2009)

19. Kittler, J., et al.: On combining classifiers. IEEE Trans. Pattern Anal. Mach. Intell. **20**(3), 226–239 (1998)

20. Lopresti, D., Zhou, J.: Using consensus sequence voting to correct OCR errors. Comput. Vis. Image Underst. **67**(1), 39–47 (1997)

21. Lund, W.B., Kennard, D.J., Ringger, E.K.: Combining multiple thresholding binarization values to improve OCR output. In: Proceedings of the International Society for Optical Engineering (SPIE) on Document Recognition and Retrieval XX. SPIE - The International Society for Optical Engineering. San Francisco, California (2013)

22. Abdulkhudhur, H.N., et al.: Implementation of improved Levenshtein algorithm for spelling correction word candidate list generation. J. Theor. Appl. Inf. Technol. **88**(3), 449–455 (2016)

Automatic Fingerprint Classification Scheme by Using Template Matching with New Set of Singular Point-Based Features

Alaa Ahmed Abbood[1]([✉]), Ghazali Sulong[2], Nada Mahdi Kaittan[1],
and Sabine U. Peters[3]

[1] Faculty of Business Informatics,
University of Information Technology and Communications (UOITC),
Baghdad, Iraq
{aaalaa2,nadait.2016}@uoitc.edu.iq
[2] School of Information and Applied Mathematics,
Universiti Malaysia Terengganu, 21030 Kuala Terengganu, Malaysia
ghazali.s@umt.edu.my
[3] College of Education, Florida State University, Tallahassee, FL 32306, USA
supeters67@gmail.com

Abstract. Fingerprint classification is used to assign fingerprints into five established classes, namely, Whorl, Left loop, Right loop, Arch, and Tented Arch, on the basis of ridge structures and singular points' trait. Although some progresses have been achieved in improving accuracy rates, problems arise from ambiguous fingerprints, especially those with large intraclass and small interclass variations. Poor-quality images, such as those with blur, dry, wet, low contrast, cut, scarred and smudgy features, are equally challenging. This study proposes a new classification technique based on template matching by using fingerprint salient features as a matching tool. In classification phase, a new set of fingerprint features is created based on singular points' occurrence and location along the symmetric axis. A set of five templates, in which each template represents a specific true class, is then generated. Finally, classification is performed by calculating similarity between the query fingerprint image and the template images by using ×2 distance measure. The performance of the current method is evaluated in terms of accuracy by using 27,000 fingerprint images acquired from the National Institute of Standard and Technology Special Database 14, which is a de facto dataset for development and testing of fingerprint classification systems. Experimental results have a high accuracy rate of 93.05%.

Keywords: Fingerprint classifications · Fingerprint features
Fingerprint image enhancements

1 Introduction

Biometrics are measurable characteristics based on physiological and behavioral traits that identify each individual. The most important type of human biometrics is fingerprints. Fingerprints are used for personal recognition in forensics, such as criminal

© Springer Nature Switzerland AG 2018
S. O. Al-mamory et al. (Eds.): NTICT 2018, CCIS 938, pp. 226–239, 2018.
https://doi.org/10.1007/978-3-030-01653-1_14

investigation tools, and civilian fields. Fingerprints are also utilized in border access control systems, national identity card validation, and authentication processors. The uniqueness and immutability of fingerprint patterns and the low cost of associated biometric equipment make fingerprints more desirable than the other types of biometrics [1, 2]. Schaeuble (1932) and Babler (1991) proved that the fingerprints of twins sharing similar DNAs are different [3].

In general, fingerprint-based recognition systems work in two modes: verification and identification. In verification mode, the systems verify the person's identity by using a 1:N comparison between the person's fingerprints and those stored in the record. Verification confirms whether the identity of the person with the fingerprint is valid. However, the process used in fingerprint identification systems is more complex than that in print verification especially for large databases because the former requires the input fingerprints to be compared with all the fingerprints in the database to find a match. Verification uses 1:1 comparison for matching, and fingerprint identification requires 1:N comparison to determine if the individual is present in the database [4, 5]. In fingerprint identification, matching accuracy and processing time are critical issues. Fingerprints in the database are organized into a number of mutually exclusive classes that share certain similar properties to efficiently identify a fingerprint. This process is called fingerprint classification [6, 7], which is the most important part of an Automatic Fingerprint Identification System (AFIS) because it provides an indexing mechanism and facilitates matching with large databases. When a class of a query fingerprint is known, matching the fingerprint requires that the print is compared with a similar class of prints. Evidence suggests that people in ancient times are aware of fingerprints. However, the full potential of fingerprints as a means of personal identification has not been realized [8]. Sir Francis Galton (1892) is the first person to study fingerprint-based identification; his work led to the first formally recognized system for fingerprint classification. Galton's classification was introduced as a means of indexing fingerprints to facilitate the search for a particular fingerprint within a collection of many prints and proposed three basic fingerprint classes, namely, Arch, Loop, and Whorl. Galton also conducted the first study on the uniqueness of fingerprints. In addition to permanence, uniqueness is necessary for a fingerprint to be a viable tool for personal identification.

Basing on Galton's work, Edward Henry (1990) subdivided two of the three main classes into specific subclasses. Henry distinguished between Arch, Tented-arch, Left Loop, Right Loop, and Whorl, as shown in Fig. 2. He also introduced the concept of fingerprint "Core" and "Delta" points and used them as aids for fingerprint classification. Henry's classification scheme constitutes the basis for most modern classification schemes [8, 9] (Fig. 1).

Fig. 1. An example of Henry's five classes [8].

2 Methodology

Fingerprint classification is a process where fingerprints are grouped in a consistent and reliable way, such that different impressions of the same fingerprint fall within the same group. In fingerprint identification systems, classification aims to speed up the database search by comparing a query fingerprint with only a small subset of fingerprints in the database belonging to the same group. Manual (human expert) classification is based essentially on the orientation of the dominant ridge flow and the presence and location of singular points (core and delta) [1]. In general, the most popular classes known as the Henry classes are the Right Loop class, Left Loop class, Tented-arch Whorl class, and Arch class.

The Right Loop class, as shown in Fig. 2(a), describes a case where some ridge pattern of a fingerprint image follows an upward flow in the right direction, then makes a curve in the center of the fingerprint ridge pattern that forms a Top Core point after which it flows out in a downward direction. The Right Loop class is also known to be described by a Delta point that forms at the opposite direction of the Right Loop, which in this case is a left side delta. The Left Loop class, as the name indicates, is the opposite of the Right Loop class and can be seen in Fig. 2(b). Different from the Right/Left Loop class is the Tented-arch class (Fig. 2c). Tented-arch class describes the case where an upward flow occurs from one side of the fingerprint ridge pattern that rises to the middle and forms a Top Core point, then flows out through the other side. At the point directly below the Core point, a Delta point is formed. For the Whorl class, the fingerprint ridges form a cycle-like pattern at the center of the fingerprint that forms two cores, namely, the Top Core formed at the upper points of the curve and the Bottom Core at the lower points of the curve. As shown in Fig. 2(d), the Whorl class is also described by two delta points that are located at both the left and right sides below the cycle-like pattern. The Arch class describes a case where the ridge pattern of a fingerprint flows from one side, raises slightly high in the middle, and then ends at the other side without forming a Core or Delta point. The Arch class can be shown in Fig. 2(e).

A number of research works have been conducted in automatic fingerprint classification, which can be categorized as ruled-based classification techniques [21] (Chong et al. 1997; Wang et al.; Maltoni 2009; Saparuldin 2012; Cao et al. 2013) and neural

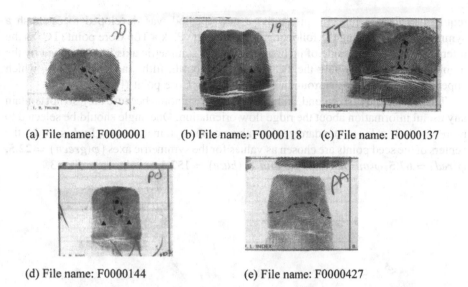

(a) File name: F0000001   (b) File name: F0000118   (c) File name: F0000137

(d) File name: F0000144   (e) File name: F0000427

Fig. 2. Ridge flow on different classes (a) Right Loop, (b) Left Loop, (c) Tented-arch, (d) Whorl (e) Arch.

network-based classification techniques [22]. However, the lack of number of features used for classification is evident, resulting in a significant number of misclassifications. To address the misclassification problem, this study utilizes several numbers of features, which invariably will result in an accurate and reliable fingerprint classification system. The classification technique used in this study involves two phases: (1) feature extraction, which aims to extract a set of distinguishable features for each class from the fingerprint images; and (2) class assignment by using Template Matching technique. This technique is developed based on five templates, which represent all the established classes. A series of experiments were conducted to find the optimal image sample from each class to build a template. The classification is performed by comparing the distance between the query fingerprint images with these templates. At the end, the query fingerprint images will be assigned into one of the five known classes: Whorl (W), Right Loop (RL), Left Loop (LL), Arch (A), and Tented-arch (TA).

2.1 Feature Extraction for Classification

The current feature extraction consists of two steps, namely, symmetric axis calculation and (2) feature selection, to find salient features that have low intraclass variation and high interclass variation.

Symmetric Axis Calculation. The local ridge flow surrounding the Top Core is very critical in class determination. Therefore, the main step in feature extraction is the calculation of symmetric axis, which allows the selection of additional features. This study begins by calculating a hypothetical line that is locally symmetric to the ridge structures at the Top Core point. Mathematically, a specific point and an angle are required to form a line. In this study, a new method was developed to establish a symmetric axis by using the following procedure. (1) Pick a Top Core point (TC) as the referral point. (2) The angle of inclination φ of the symmetric axis is determined by the dominant color surrounding the TC in the outer fourth, fifth, and sixth blocks, which appear as square shapes surrounding a print's Top Core point (Fig. 3).

The color in the first, second, and third blocks are ignored because they do not contain any useful information about the ridge flow orientation. One angle should be selected to represent each color to standardize the values of the symmetric axes. In this study, the centers of the seed points are chosen as values for the symmetric axes ($\varphi(green) = 22.5$, $\varphi(red) = 67.5$, $\varphi(purpil) = 112.5$ and φ $(blue) = 157.5$), as shown in Fig. 3.

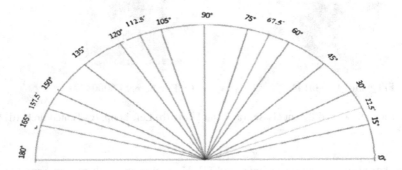

Fig. 3. Shows the four symmetric axis angles φ according to colour. (Color figure online)

Figure 4 illustrates an example of this process. The angle of the symmetric axis φ is determined by calculating the dominant colors in the extreme fourth, fifth, and sixth blocks surrounding the TC. The value for φ was assigned the same value similar to that of the dominant color. The diagram shows that the blue (B) color has 39 blocks, and the red (R), green (G), and purple (P) have 24, 27, and 30 blocks, respectively. As a result, the symmetric axis has an angle value of 157.5°, which is representative of the center of the blue color seed.

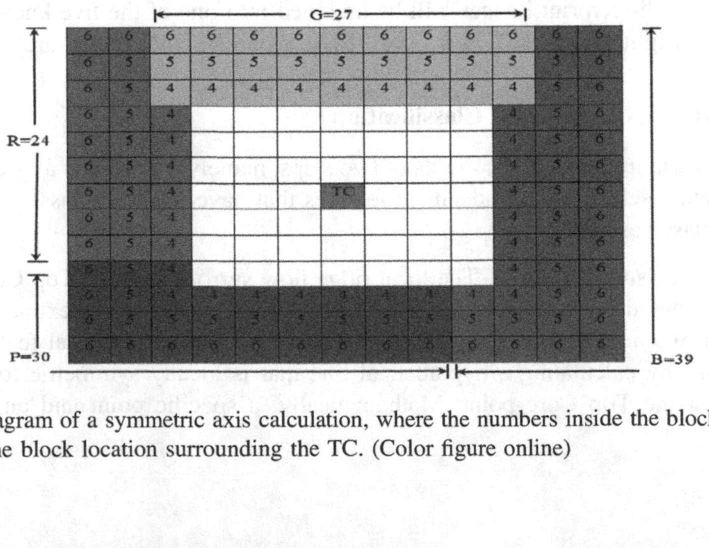

Fig. 4. Diagram of a symmetric axis calculation, where the numbers inside the blocks (4, 5, 6) represent the block location surrounding the TC. (Color figure online)

Classification Features. Feature extraction is aimed at selecting the maximum number of features from an image. In this study, nine features were identified $[f_1, ..., f_9]$ based on the type and position of singularities and the location of these singularities along a symmetric axis. Given that the symmetric axis follows the ridges flow, these features are invariant to the scaling and rotation and are described as follows:

f_1 : If (the image contained a Top Core) then $f_1 = 1$, else 0.

f_2: If (the image contained a Bottom Core) then $f_2 = 1$, else 0.

f_3: If (the image contained a Delta) then $f_3 = 1$, else 0.

f_4: If (the image contained a Second Delta) then $f_4 = 1$, else 0.

f_5: If ($f_1 = 0$ or $f_3 = 0$) then $f_5 = 0$.
If (a Delta is on the right side of the symmetric axis) then $f_5 = 1$
Else If (a Delta is on the left of the symmetric axis) then $f_5 = 4$

f_6: If ($f_1 = 0$ or $f_4 = 0$) then $f_6 = 0$
If (the 2nd. Delta is on the right side of the symmetric axis) then $f_6 = 1$.
Else If (the 2nd. Delta is on the left of the symmetric axis) then $f_6 = 2$.

f_7: If ($f_1 = 0$ or $f_3 = 0$) then $f_7 = 0$.
If (the perpendicular distance between a Delta and the symmetric axis less than 4 blocks) then $f_7 = 1$.
Else If (the perpendicular distance between a Delta and the symmetric axis not lesser than 4 blocks) then $f_7 = 2$

f_8: If ($f_1 = 0$ or $f_4 = 0$) then $f_8 = 0$
If (the perpendicular distance between a second Delta and the symmetric axis lesser than 4 blocks) then $f_8 = 1$.
Else If (the perpendicular distance between a second Delta and the symmetric axis not lesser than 4 blocks) then $f_8 = 2$.

f_9: If ($f_1 = 0$) then $f_9 = 0$.
If (symmetric axis angle equalled, $22.5°$ or $67.5°$) then $f_9 = 1$.
Else If (symmetric axis angle equalled $112.5°$ or $157.5°$) then $f_9 = 2$.

All the above features are used for fingerprint classification. Figure 5 shows an example of feature extraction on the fingerprint Whorl class.

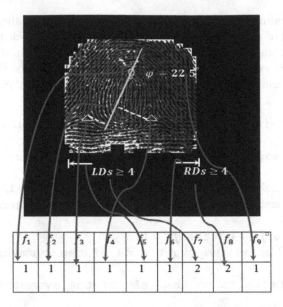

Fig. 5. Anexample of feature extraction on the fingerprint Whorl class.

2.2   Fingerprint Classification Using Template Matching Technique

Once the features of each fingerprint image have been extracted, the exclusive class where each fingerprint belongs to is determined. However, classification becomes difficult due to the existence of imperfect features for some fingerprint images. For this reason, creating templates for each fingerprint class to compare a query print and the template is important.

Image Template Selection. Atemplate for each print class was created using 40 images arbitrarily selected from the dataset. The images in the dataset have already been classified into true and non-ambiguous classes (non-ambiguous means that the images do not have properties of more than one class) by human experts. The choice of the number of images chosen in each template is based on the numerous experiments run on a subset of the dataset that consists of 1,000 images. For this subset of the dataset, classification is carried out by considering five images for each class as a template. Subsequently, an increase by five to the number of images in the template is made at each time of the classification until it reaches a saturation point. The result of this experiment is illustrated in Fig. 6.

From the figure, the accuracy of the classification increases at a fast rate as the number of images increases from 5 to 40; however, a slow increase or perhaps stagnant in accuracy was observed from the point where the number of images selected reached 40 and more. According to Bien and Tibshirani (2011), the minimum number of template that achieves high classification accuracy is efficient for template-based classification. Therefore, in this study, the number of templates chosen per class is 40, which is based on the high classification accuracy for this number of template observed in Fig. 9.

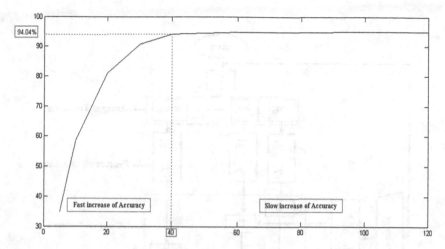

Fig. 6. Illustrate the process of selecting the optimum number of image template for each class.

Fingerprint Class Assignment. Once the image templates are selected for each class C_i, the next process is to assign the query image into its exclusive class. This process starts by extracting the features for template images and query image I. The distances between the respective fingerprints' features are computed using x^2 Distance measure by Eq. (1).

$$D(p,q) = \sum_{i=1}^{9} \frac{(p_i - q_i)^2}{p_i + q_i} \tag{1}$$

$$if\,(p_i = 0\ and\ q_i = 0)\ then \frac{(p_i - q_i)^2}{p_i + q_i} is\ set\ to\ 0$$

Where, p_i represents the feature vectors of the template images, and q_i are the feature vectors of the query image.

If the distance D between the respective features is equal to 0, then the query image I is assigned to the class C_i that has the value 0. Otherwise, if $D \neq 0$ for all the 40 images of the class i, then the average distance $\bar{D}$ is calculated. This process is continued by increasing the value of i by one step until the five classes are obtained. If all of the calculated average values are not zero, then the query image will be assigned to the class that have the minimum average value. For a clear illustration for the abovementioned processes, a flowchart is presented in Fig. 7.

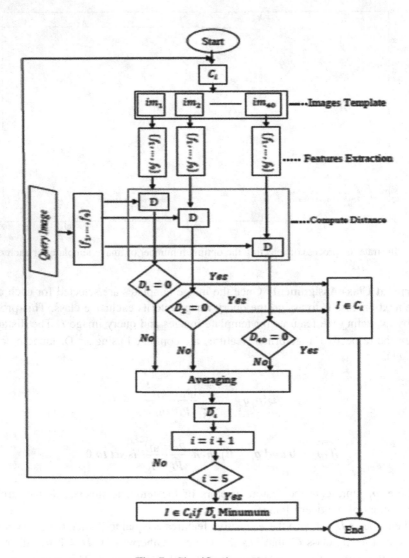

Fig. 7. Classification process.

The pseudocode of the classification is given in Algorithm 1 below.

```
Input: Database F₁, F₂, ... ... F₂₇₀₀₀ List of FingerPrint image.
       {Fᵢ = {fᵢ₁, fᵢ₂ ... ... fᵢ₉}. Features of image i.
Output: Classified fingerprints according to their class.
    Begin:
          for j= 1→ 5, increment by 1   do
              Put 40 arbitrary images from class j in the List Cⱼ.
          end for
          for i= 1→ 27000, increment by 1   do
             MinDis ← +∞;
             Index  ← -1;
             for j= 1 → 5, increment by 1  do
                SumDis ← 0;

                   for k = 1→ 40, increment by 1 do
                     Disₖ ← x²_Distance (Fᵢ, Fₖ).
                     If ( Disₖ = 0 ) then
                          Assign Image Fᵢ to the class Cⱼ;
                          Put the Image Fᵢ in the List Cⱼ;
                          Break the loop;
                     else
                          SumDis ← SumDis + Disₖ;
                     end if
                   end for
                AvrgDis ← SumDis  / 40;
                If (AvrgDis < MinDis )   then
                      MinDis ← AvrgDis;
                      Index ← j;
                end If
             end for
             If Index    != -1
                Assign Image Fᵢ to the class C_Index .
                Put the Image Fᵢ in the List C_Index .
             end If
          end for
    End
```

3   Results and Discussion

The performance of a fingerprint classification system is usually measured in terms of error rate or accuracy. The error rate is computed as the ratio between the number of misclassified fingerprints and the total number of samples in the dataset (Eq. (2)). The accuracy is the percentage of correctly classified fingerprints and can be calculated using Eq. (3) below.

$$Error\ rate = \frac{Number\ of\ misclassified\ fingerprints \times 100}{total\ number\ of\ fingerprint} \tag{2}$$

$$Accurcy = 100 - error\ rate \tag{3}$$

A confusion matrix is created based on the above error rate that will be used to obtain further information about the classifier behavior for each class. This matrix has a row for each true class (ground truth), which has been classified according to the human expert assigned by FBI, and is labeled in NIST 14 and a column for each hypothesized class obtained from the proposed classification method. Each cell at row r and column c reports how many fingerprints are belonging to class r are assigned to class c.

The experimental results compiled in the confusion matrix are presented in Table 1; the current classification technique classified all the 27,000 samples with total accuracy of 93.05% on the correct class detection, and 6.78% was misclassified. Based on the confusion matrix, the results also proved that the current method has markedly outperformed the results of the classification techniques proposed by [1, 23] on the same dataset (Fig. 8), where the total accuracies were 81.2% and 89.31%, respectively.

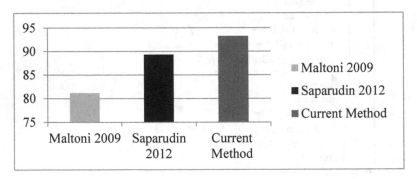

Fig. 8. A comparison in total accuracy between the current method and previous studies on same dataset.

Table 1 shows the results of the current classification technique in the form of a confusion matrix. Based on the hypothesized classes, the outcome of the current method was five classes, whereas Saparudin's method came up with six classes (including the Unknown class) because some images are not matched with any of the rules that were applied.

$$Error\ rate = \frac{1877 \times 100}{27000} = 6.95 \tag{4}$$

Table 1. A confusion matrix of the experimental results for the current fingerprint classification technique.

True class	Hypothesized class from the proposed method				
	Whorl	Left-loop	Right-loop	Arch	Tented-arch
Whorl	8098	42	52	21	117
Left-loop	89	7987	373	160	12
Right-loop	79	389	7500	167	105
Arch	8	47	22	856	43
Tented-arch	14	64	57	16	664
Scars	0	1	6	7	7

$$Accurcy = 100 - 6.95 = 93.05 \tag{5}$$

The current method also improved the individual accuracy of each class to 97.21%, 92.64%, 91.01%, 88.06%, and 81.37% for Whorl, Left-loop, Right-loop, Arch, and Tented-arch, respectively. Meanwhile, the individual accuracies of each class based on Saparudin's classification technique were 87.90%, 91.02%, 90.20%, 86.61%, and 79.98% for Whorl, Left-loop, Right-loop, Arch, and Tented-arch, respectively [23]. The current classification technique significantly improves all classes, especially Whorl, which consists more available features than the other classes. Therefore, missing some of these features will not affect the classification of this print. Notably, the above accuracy percentages achieved excluded the scar images. The comparison between results in terms of individual accuracy for both techniques is illustrated in Fig. 9.

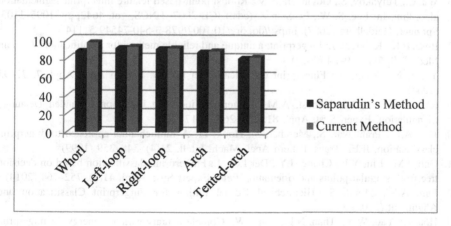

Fig. 9. Comparison of the current method with Saparudin's one for each individual class.

4 Conclusion

Experiments were conducted using NIST Special Database 14 with 27,000 sample sizes for testing the performance of the current classification technique. The classification used in this study was tested and benchmarked with the latest study of Maltoni (2009) and Saparudin [23]. Overall, the current classification method has precisely classified 93.05% of the 27,000 fingerprints, which significantly outperformed the latest work of Saparudin [23] and Cappelli and Maltoni (2009), whose research's activities are mainly focused on fingerprints and are considered the most respected researchers in this field. Interestingly, most fingerprints including good, dry, wet, bruises, stains, cuts, and low contrast are precisely classified. By contrast, only small percentage of the NIST 14 fingerprints is misclassified and includes small-interclass variation, ambiguity, and poor-quality fingerprint images.

References

1. Maltoni, D., Cappelli, R.: Advances in fingerprint modeling. Image Vis. Comput. **27**(3), 258–268 (2009)
2. Ferrara, M., Cappelli, R., Maltoni, D.: On the feasibility of creating double-identity fingerprints. IEEE Trans. Inf. Forensics Secur. **12**(4), 892–900 (2017)
3. Boxwala, M., Boxwala, M., Jasra, P.K.: Analysis of ridge characteristics of fingerprint from different fingers of monozygotic twins. J. Emerg. Forensics Sci. Res. **1**, 78–82 (2016)
4. Maltoni, D.: A tutorial on fingerprint recognition. In: Tistarelli, M., Bigun, J., Grosso, E. (eds.) Advanced Studies in Biometrics. LNCS, vol. 3161, pp. 43–68. Springer, Heidelberg (2005). https://doi.org/10.1007/11493648_3
5. Galar, M., et al.: A survey of fingerprint classification Part II: experimental analysis and ensemble proposal. Knowl. Based Syst. **81**, 98–116 (2015)
6. Wu, C., Tulyakov, S., Govindaraju, V.: Robust point-based feature fingerprint segmentation algorithm. In: Lee, S.-W., Li, Stan Z. (eds.) ICB 2007. LNCS, vol. 4642, pp. 1095–1103. Springer, Heidelberg (2007). https://doi.org/10.1007/978-3-540-74549-5_114
7. Bose, P.K., Kabir, M.J.: Fingerprint: a unique and reliable method for identification. J. Enam Med. Coll. **7**(1), 29–34 (2017)
8. Yager, N., Amin, A.: Fingerprint classification: a review. Pattern Anal. Appl. **7**, 77–93 (2004)
9. Khodadoust, J., Khodadoust, A.M.: Fingerprint indexing based on expanded Delaunay triangulation. Expert Syst. Appl. **81**, 251–267 (2017)
10. Jain, A.K., Prabhakar, S., Member, S., Hong, L.: A multichannel approach to fingerprint classification. IEEE Trans. Pattern Anal. Mach. Intell. **21**(4), 348–359 (1999)
11. Guo, J.M., Liu, Y.F., Chang, J.Y., Der Lee, J.: Fingerprint classification based on decision tree from singular points and orientation field. Expert Syst. Appl. **41**(2), 752–764 (2014)
12. Jain, A.K., Minut, S.: Hierarchical Kernel Fitting for Fingerprint Classification and Alignment (2002)
13. Hou, Z., Yau, W.Y., Than, N.L., Tang, W.: Complementary variance energy for fingerprint segmentation. In: Satoh, S., Nack, F., Etoh, M. (eds.) MMM 2008. LNCS, vol. 4903, pp. 113–122. Springer, Heidelberg (2008). https://doi.org/10.1007/978-3-540-77409-9_11

14. Wang, X., Xie, M.: Fingerprint classification: an approach based on singularities and analysis of fingerprint structure. In: Zhang, D., Jain, A.K. (eds.) ICBA 2004. LNCS, vol. 3072, pp. 324–329. Springer, Heidelberg (2004). https://doi.org/10.1007/978-3-540-25948-0_45

15. Dorasamy, K., Webb, L., Tapamo, J., Khanyile, N.P.: Fingerprint classification using a simplified rule-set based on directional patterns and singularity features. In: Proceedings of the 2015 International Conference on Biometrics, ICB 2015, vol. 2, pp. 400–407 (2015)

16. Peralta, D., Triguero, I., García, S., Saeys, Y., Benitez, J.M., Herrera, F.: Distributed incremental fingerprint identification with reduced database penetration rate using a hierarchical classification based on feature fusion and selection. Knowl. Based Syst. **126**, 91–103 (2017)

17. Bian, W., Ding, S., Xue, Y.: Combining weighted linear project analysis with orientation diffusion for fingerprint orientation field reconstruction. Inf. Sci. (Ny) **396**, 55–71 (2017)

18. Kumar, G.S.R., et al. (eds.): Proceedings of the Second International Conference on Soft Computing for Problem Solving (SocProS 2012), December 28-30, 2012. Springer, Cham (2014)

19. Li, Y., Mandal, M., Lu, C.: Singular point detection based on orientation filed regularization and poincaré index in fingerprint images. In: 2013 IEEE International Conference on Acoustics, Speech and Signal Processing, pp. 1439–1443 (2013)

20. Hsieh, C., Shyu, S., Hung, K.: An effective method for fingerprint classification. Tamkang J. Sci. Eng. **12**(2), 169–182 (2009)

21. Blue, J.L., Candela, G.T., Grother, P.J., Chellappa, R., Wilson, C.L.: Evaluation of pattern classifiers for fingerprint and OCR applications. Pattern Recognit. **27**(4), 485–501 (1994)

22. Assas, O., Aijimi, A., Boudrah, I., Bouamar, M.: Comparison of neuro-fuzzy networks for classification fingerprint images. J. Comput. Eng. Inf. Technol. 1–5 (2013)

23. Saparudin: An Automatic Fingerprint Classification Technique Based on Singular Points and Structure Shape of Orientation Fields. Ph.D. thesis (2012)

Human Skin Colour Detection Using Bayesian Rough Decision Tree

Ayad R. Abbas and Ayat O. Farooq$^{(\boxtimes)}$

Department of Computer Science, University of Technology, Baghdad, Iraq
ayad_cs@yahoo.com, ayatomar133@yahoo.com

Abstract. Human skin colour detection is a necessary element in computer vision as well as in image processing applications. It is a separation procedure of skin pixels from non-skin pixels in image. Whereas, detecting human skin colour is a very difficult function due to two reasons, mainly, changeable illumination conditions and diverse races of people. Some previous researchers in this field tried to resolve those problems by using thresholds that relied on specific values of skin tones. Although, it is a speedy and an easy implementation, it does not provide sufficient information for recognizing all skin tones of humans. This paper proposes Bayesian Rough Decision Tree (BRDT) classifier to improve the accuracy of human skin detection. Three experiments have been conducted using (RGB) dataset collected from University of California, Irvine (UCI) machine learning repository, RGB (Red, Green, Blue), HSV (Hue, Saturation, Value) and YCbCr (Luminance, Chrominance). The experimental result shows that the proposed system can achieve preferable accuracy in skin detection 98%, 97% and 97% using RGB dataset, HSV dataset and YCbCr dataset respectively.

Keywords: Skin detection · Machine learning · Bayesian rough set
Decision tree

1 Introduction

The colour of human skin is constructed by a mixture of blood red colour and melanin which is composed of brown and yellow colours. The fundamental aim of human skin colour detection is to construct understandable rules that will differentiate between skin and non-skin pixels in image [1]. Human skin colour detection has a substantial role in image processing as well as in computer vision applications. It is a preprocessing step in diverse applications such as hand and face detection, personal identification, security systems, human computer interaction, intelligent video surveillances and skin diseases [2, 3].

Although skin colour is invariant to orientation and rotation, detection procedure of it is a very difficult function because it is sensitive to four problems that contribute mainly and directly to changeable skin colour in that image like [1]:

1. Illumination factor: Any variation in the degree of image illumination whether bright illumination or dark illumination will affect significantly skin colour in the image.

© Springer Nature Switzerland AG 2018
S. O. Al-mamory et al. (Eds.): NTICT 2018, CCIS 938, pp. 240–254, 2018.
https://doi.org/10.1007/978-3-030-01653-1_15

2. Specifications of the camera: skin colours in images rely on specifications of the camera. Even if images are captured under the same circumstances, they will differ from one camera to another according to the specifications of the camera sensor.
3. Ethnic groups: There are varying ethnic groups of people such as Asian, African, Caucasian and Hispanic; these groups vary from one to another, in skin colour tones, ranging from white, yellow, to dark colours.
4. Characteristics of individuals: Age, sex and different body parts of people, highly contribute to changeable appearances of skin colours.

This paper focused on resolving the above problems by applying a multiple of machine learning techniques: Bayesian Decision Tree, in order to classify two classes of skin colour pixel and non-skin colour pixel. The essential goal of our proposed system being improving the performance of skin detection procedure in order to achieve preferable accuracy in detection, using skin segmentation dataset picked from (UCI) machine learning repository [4].

2 Related Work

There are numerous previous researches on human skin colour detection as is shown below:

Fuzzy System - Fuzzy Clustering and Support Vector Machine [5]: The hybrid techniques aims at connecting advantages and removing disadvantages of both Fuzzy neural network and support vector machine. They were worked on the dataset where total of 450 images, which includes 27 various people of different races and facial expressions. Each image consists of resolution 896 ×592 pixels.

Fuzzy Decision Tree [6]: Fuzzy Decision Tree detects skin in images, in a few seconds. It is highly effective for application into embedded devices. To determine the timing performance, all the images had been scaled to standard 640 * 480 resolution.

A threshold-based algorithm [7]: The technique differentiates skin images from non-skin images using the RGB, HSV and YCbCr. The technique is capable of processing images under various illumination conditions such as brightness.

Fuzzy System [8]: This is used for detection of skin colour in RGB images. It is extracting only the skin information, using different ethnic groups and variable lighting conditions.

Adaptive Neuro Fuzzy Inference Systems [9]: These are used for detection of skin from non-skin. This hybrid technique is capable of detecting skin pixels by picking the same dataset of RGB of both skin and non-skin pixels that is used in our work.

Although all techniques mentioned above, used by previous researchers in this field can achieve good results, our proposed system can achieve better accuracy 98% in human skin colour detection procedure for variation of races of people and under variant illumination conditions with resolution 256 × 256 pixel. Moreover, this resolution was used to reduce the time complexity by applying only discretization method on RGB image or HSV image, however both normalization and discretization methods are applied on YCbCr image.

3 Skin Segmentation Dataset (RGB) Description

In this paper, skin segmentation dataset (RGB) is used which was collected from UCI machine learning repository [4]. This dataset is picked by using skin textures of various face images of people of different ages, gender, and race groups, and non-skin pixels obtained from arbitrary thousands of random sampling of textures. RGB dataset is composed of three attributes information, each of which attributes appears as an integer value between 0 to 255 with two classes, 1 for skin and 2 for non-skin. Total learning sample size is 245057, the number of skin sample is 50859 and the number of non-skin sample is 194198 as shown in Table 1.

Table 1. The skin segmentation dataset

Domain	R	G	B	Class
U1	123	85	74	1
U2	123	177	178	2
⋮	⋮	⋮	⋮	⋮
U245057	255	255	255	2

4 HSV Dataset Description

HSV colour space is more axiomatic to how people experience colour than the RGB colour space. As hue (H) varies from 0 to 1.0, the corresponding colours differ from red, through yellow, green, cyan, blue, and magenta, back to red. As saturation (S) differs from 0 to 1.0, the corresponding colours (hues) differ from unsaturated (shades of grey) to fully saturated (no white component) [7]. HSV dataset is obtained by applying Eq. (1) for each value of conditional attributes of RGB dataset [10], HSV dataset can be described in Table 2.

$$\begin{cases} H = \arccos \dfrac{\frac{1}{2}(2R-G-B)}{\sqrt{(R-G)^2 + (R-B)(G-B)}} \\ S = \dfrac{max(R,G,B)-min(R,G,B)}{max(R,G,B)} \\ V = max(R,G,B) \end{cases} \tag{1}$$

Table 2. The HSV dataset of skin detection

Domain	H	S	V	Class
U1	0.037415	0.398374	123	1
U2	0.503268	0.312883	178	2
⋮	⋮	⋮	⋮	⋮
U245057	0	0	255	2

5 YCbCr Dataset Description

YCbCr is an encoded non-linear RGB signal, used by European television studios and for image compression work [11]. The separation of luminance and chrominance components makes YCbCr colour space [12]. Luminance information is stored as a single component (Y), and chrominance information is stored as two colour-difference components (Cb and Cr). Cb means the difference between the blue component and reference value. Cr means the difference between the red component and a reference value [7]. YCbCr dataset is obtained by applying Eq. (2) for each value of conditional attributes of RGB dataset [10], YCbCr dataset can be described in Table 3.

$$\begin{bmatrix} Y \\ Cb \\ Cr \end{bmatrix} = \begin{bmatrix} 16 \\ 128 \\ 128 \end{bmatrix} + \begin{bmatrix} 0.279 & 0.504 & 0.098 \\ -0.148 & -0.291 & 0.439 \\ 0.439 & -0.368 & -0.071 \end{bmatrix} \begin{bmatrix} R \\ G \\ B \end{bmatrix} \tag{2}$$

Table 3. The YCbCr dataset of skin detection

Domain	Y	Cb	Cr	Class
U1	81.74373	− 9.96191	17.97786	1
U2	126.4507	8.352353	− 21.5303	2
⋮	⋮	⋮	⋮	⋮
U245057	219.0627	0.501961	0.501961	2

6 Data Preprocessing

Dataset in real world may contains noise, missing values and inconsistency. The quality of dataset affects the data mining results. Data preprocessing technique which depicts any sort of processing, performed on raw dataset, to equip it for another analyzing procedure. Data preprocessing techniques reconstruct the dataset into a format that will be better understood by the user, and more efficient for further processing [13].

6.1 Discretization

Discretization is one of the data preprocessing techniques that converts continuous values into interval. Discretization can perform recursive procedure on every conditional attribute in dataset. The substantial part of discretization procedure is selecting the preferable cut off points which partition the continual value range into separate number of bins [14].

6.1.1 Equal Frequency

Equal frequency technique has been used to preprocess continual values in conditional attributes, into separate values, that are preformed to evaluate decision tree classifiers, through many experiments, that construct the optimal Equal Frequency discretization technique which leads up to high quality discretization [15].

In this paper, each continuous value from 0 to 255 in RGB and V attributes are converted to interval [0–19] using equal frequency discretization technique as shown in Table 4 while each continuous value of H and S attributes from 0 to 1 are converted to interval [0–19] as shown in Table 5.

Table 4. The equal frequency discretization of RGB dataset

Bin	Range
1	[0–29]
2	[30–38]
3	[39–48]
4	[49–57]
5	[58–60]
6	[61–74]
7	[75–89]
8	[90–94]
9	[95–100]
10	[101–104]
11	[105–115]
12	[116–124]
13	[125–138]
14	[139–145]
15	[146–154]
16	[155–170]
17	[171–181]
18	[181–211]
19	[212–240]
20	[241–255]

Table 5. The equal frequency discretization of HSV dataset

Bin	Range
1	[0–0.0999]
2	[0.1000–0.1499]
3	[0.1500–0.1999]
4	[0.2000–0.2499]
5	[0.2500–0.2999]
6	[0.3000–0.3499]
7	[0.3500–0.3999]
8	[0.4000–0.4499]
9	[0.4500–0.4999]
10	[0.5000–0.5499]

(continued)

Table 5. (*continued*)

Bin	Range
11	[0.5500–0.5999]
12	[0.6000–0.6499]
13	[0.6500–0.6999]
14	[0.7000–0.7499]
15	[0.7500–0.7999]
16	[0.8000–0.8499]
17	[0.8500–0.8999]
18	[0.9000–0.9499]
19	[0.9500–0.9999]
20	[1]

Table 6. The equal frequency discretization of YCbCr dataset

Bin	Range
1	[0–0.0999]
2	[0.1000–0.1099]
3	[0.1200–0.1299]
4	[0.1300–0.1499]
5	[0.1500–0.1699]
6	[0.1700–0.1899]
7	[0.2000–0.2099]
8	[0.2200–0.2299]
9	[0.2300–0.2499]
10	[0.2500–0.2699]
11	[0.2700–0.2899]
12	[0.3000–0.3099]
13	[0.3200–0.3299]
14	[0.3300–0.3499]
15	[0.3500–0.3699]
16	[0.3700–0.3899]
17	[0.4000–0.4099]
18	[0.4200–0.4299]
19	[0.4300–0.4499]
20	[0.4500–1]

6.2   Normalization

Normalization is a data preprocessing tool utilized in data mining system. An attribute of a dataset is normalized by scaling its values so that they fall within a little-specified range, such as 0.0–1.0. Normalization is particularly useful for many classification algorithms involving neural networks. There are many methods for data normalization such as min-max normalization, z-score normalization and normalization by decimal scaling.

Min-Max Normalization. Min-max normalization performs a linear transformation on the raw data. The min-max normalization is computed by the following Eq. (3) [16]:

$$min - max = \frac{v(A) - min(A)}{max(A) - min(A)} (newmax - newmin) + newmin \qquad (3)$$

Where: $v(A)$ is value in attribute A, $min(A)$ represents minimum value in attribute A, $max(A)$ represents maximum value in attribute A, $newmax$ is 1, $newmin$ is 0.

YCbCr attributes is normalized to range [0–1] by using min-max normalization. Since, min-max normalization method with YCbCr dataset gives low skin detection. Therefore, equal frequency discretization is applied as shown in Table 6 to improve skin detection.

The twenty bins are obtained through many experiments based on trial and error. These bins are used in order to get high accuracy in skin detection.

7 BRDT Algorithm

The proposed system is incorporating benefits of (Bayesian Rough Set [17] and Decision tree [18]) whereas, our method enhanced the performance of weakness skin detection through increasing the accuracy detection and decreasing the time complex. The proposed system can be described by the following steps:

Step 1: Determine the value of threshold (a).
Where: $0.1 \leq a \leq 0.9$
Step 2: Compute the posterior probability $P(A_i|X_j)$ and prior probability $P(X_j)$ for each conditional attribute, as shown in the Eq. (4) rather than compute entropy of decision tree.

$$P(A_i|X_j) = \frac{P(A_i \cap X_j)}{P(X_j)} \qquad (4)$$

Where: $P(X_j)$ is prior probability of decision attribute, A_i represents the value of each distinct value i in conditional attribute A, X_j represents the value of each class j in dataset.

The result of $P(A_i|X_j)$ is compared with a, if the result of $P(A_i|X_j)$ is greater than or equals to (a) then it is positive region $POS(X_j)$. If result of $p(A_i|X_j)$ is smaller than or equals to $(1 - a)$, then it is negative region $NEG(X_j)$. Otherwise, it is boundary region $BND(X_j)$, and can be expressed in Eq. (5).

$$\begin{cases} POS(X_j) = \cup \{A_i : P(A_i|X_j) > = a\} \\ NEG(X_j) = \cup \{A_i : P(A_i|X_j) \leq 1 - a \\ BND(X_j) = \cup \{A_i : 1 - a < P(A_i|X_j) < a\} \end{cases} \qquad (5)$$

Step 3: Compute the discriminant index for each conditional attribute of dataset that is defined in Eq. (6), rather than compute information gain of decision tree.

$$\eta = 1 - \frac{card(BND(X_j))}{card(U)} \qquad (6)$$

Where: η is the discriminant index, U is number of records in dataset, *card* is the cardinality of the set.

Step 4: Choose the conditional attribute that represents the highest discriminant index as the root node of tree, and then divide the dataset according to preferable discriminant index.

Step 4a: When the result of $P(A_i|X_j)$ for highest discriminant index is $POS(X_j)$ or $NEG(X_j)$ as leaf node.

Step 4b: When the result of $P(A_i|X_j)$ for highest discriminant index is neither $POS(X_j)$ nor $NEG(X_j)$, then it is $BND(X_j)$ that defines as non-leaf node. $BND(X_j)$ requires further splitting.

Step 5: The tree is run recursively on the non-leaf nodes, until all dataset is classified, then create set of rules.

The proposed system can also improve decision tree though calculating η rather than calculating information gain of decision tree. In addition, when all conditional attributes (RGB) are split and additional conditional attribute is not found in dataset, also, there is need for further splitting to distinguish skin or non-skin pixels, if the number of skin is greater than non-skin then skin is left, otherwise non-skin.

8 Skin Detection System

Figure 1 illustrates the skin detection system, which consists of three colour spaces, RGB, HSV and YCbCr, to produce better rate results for skin colour detection. The proposed BRDT is applied in order to classify and label all pixels separately and mark them in test image. In other words, if pixels are classified as skin, we label these pixels in red, otherwise non-skin pixels keep unchanged. The following colour spaces are used with learning-based BRDT algorithm:

1. RGB colour space: Firstly, the discretization method was performed using equal frequency discretization on RGB dataset, in order to reduce time complexity of skin detection. We apply BRDT on RGB training dataset and evaluate the BRDT through new test RGB image, the equal frequency discretization is applied on pixels of RGB image and these pixels are matched to prediction rules and finally detection of skin pixels of the RGB image.
2. HSV colour space: Firstly, we extract HSV colour space from RGB dataset using Eq. (1). After that and as mentioned RGB colour space, the discretization method was also performed using equal frequency discretization on HSV dataset. We also apply BRDT on HSV training dataset and evaluate the BRDT through new test RGB image is converted into HSV image using Eq. (1), the equal frequency discretization is applied on pixels of HSV image and these pixels are matched to prediction rules. Finally detection of skin pixels of the HSV image.

3. YCbCr colour space: Firstly, we extract YCbCr colour space from RGB dataset using Eq. (2). After that, this colour space is normalized to range [0–1] by using min-max normalization as shown in Eq. (3), then discretization method was also performed using equal frequency discretization on YCbCr dataset. We also apply BRDT on YCbCr training dataset and evaluate the BRDT through new test RGB image is converted into YCbCr image using Eq. (2), the equal frequency discretization and normalization are applied on pixels of image and these pixels are matched to prediction rules. Finally detection of skin pixels of the YCbCr image.

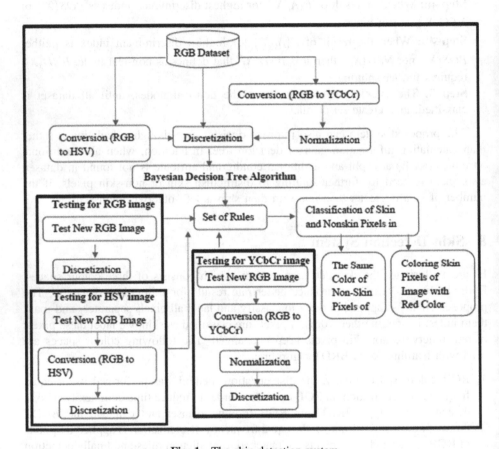

Fig. 1. The skin detection system

9   Experimental Results and Evaluation

A set of quantitative and qualitative experiment was achieved to analyze and evaluate the performance of the proposed system on human skin detection accuracy using BRDT in order to improve performance of decision tree to classify the skin pixel and non-skin pixel. Whereas, BRDT is performed with three experiments: RGB colour space, HSV colour space and YCbCr colour space.

Table 7. Results of experiment of skin detection with three experiments using BRDT

The valueof threshold	Skin detection with bright illumination	Skin detection with dark Illumination
From 0.1 to 0.5 in experiment (1)		
0.6 in experiment (1)		
0.7 in experiment (1)		
0.8 in experiment (1)		

	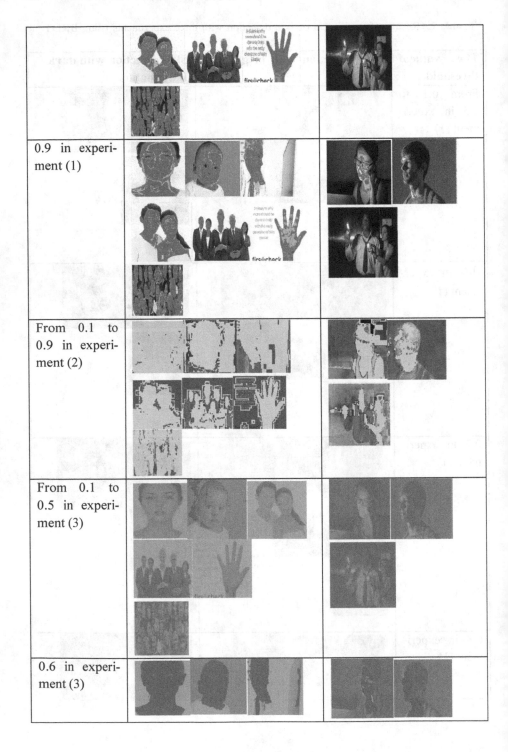	
0.9 in experiment (1)		
From 0.1 to 0.9 in experiment (2)		
From 0.1 to 0.5 in experiment (3)		
0.6 in experiment (3)		

	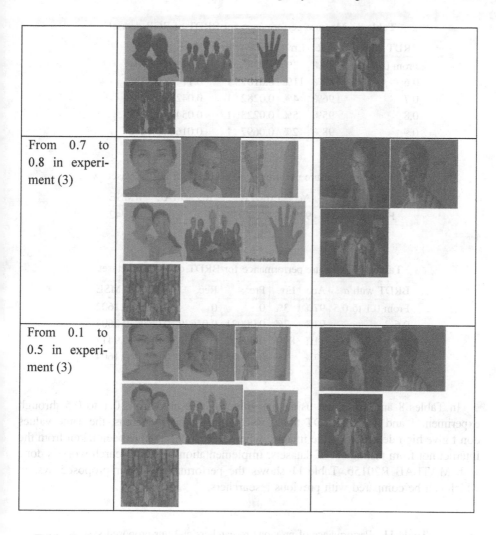	
From 0.7 to 0.8 in experiment (3)		
From 0.1 to 0.5 in experiment (3)		

Table 7, shows the value of threshold for experiments 1, 2 and 3. This indicates that high skin detection on image with bright illumination was found with experiment 1, 2 and 3 are 0.7, 0.8, from 0.1 to 0.9 and 0.6 respectively. Whereas, low skin detection was found with experiment 1 and 3 when the value of threshold from 0.1 to 0.6, 0.8, 0.9 and from 0.1 to 0.5, 0.8, 0.9 respectively. However, this indicates that high skin detection on image with dark illumination was found with experiment 2 and 3 are 0.6 and from 0.1 to 0.9 respectively. While, low skin detection was found with experiment 1 and 3 when the value of threshold from 0.1 to 0.9, and from 0.1 to 0.5 and from 0.7 to 0.9 respectively where the performance of BRDT is evaluated by splitting dataset into training set and testing set. In this paper, the training of dataset is 80% that represents 196046 of dataset and testing set of dataset is 20% that represents 49011 of dataset, however, testing set of dataset is used to calculate accuracy (Acc.), error rate (Err.), precision (Pre.), recall (Rec.), mean error absolute (MAR) and root mean square error (RMSE) as shown in Tables 8, 9 and 10.

Table 8. Measure performance for BRDT of RGB dataset

BRDT with a	Acc.	Err.	Pre.	Rec.	MAE	RMSE
From 0.1 to 0.5	97%	3%	0.0283	1	0.0290	0.1704
0.6	89%	11%	0.0107	1	0.1143	0.3381
0.7	96%	4%	0.0282	1	0.0427	0.2067
0.8	95%	5%	0.0223	1	0.0544	0.2333
0.9	98%	2%	0.0697	1	0.0166	0.1288

Table 9. Measure performance for BRDT of HSV dataset

BRDT with a	Acc.	Err.	Pre.	Rec.	MAE	RMSE
From 0.1 to 0.9	97%	3%	0.0393	1	0.0304	0.1742

Table 10. Measure performance for BRDT of YCbCr dataset

BRDT with a	Acc.	Err.	Pre.	Rec.	MAE	RMSE
From 0.1 to 0.5	97%	3%	0	0	0.0257	0.1602
0.6	93%	7%	0.00031	0.01639	0.0678	0.2604
From 0.7 to 0.8	11%	89%	0.00002	0.01639	0.8868	0.9417
0.9	97%	3%	0.00083	0.00002	0.0257	0.1604

In Tables 8 and 10, when using the value of threshold from 0.1 to 0.5 through experiment 1 and 3 with BRDT that gives high accuracy, whereas, the same values don't give high detection on the images, where these images have been taken from the Internet not from testing set of datasets. Implementation of this research work is done with MATLAB R2015b. Table 11 shows the performance of our proposed system which can be compared with previous researchers.

Table 11. Performance of previous researchers and our proposed system

Machine learning methods	No. of training datasets	Acc.
Adaptive Neuro Fuzzy Inference Systems	245057 pixels	(89.40%–90.10%)
Bayesian Decision Tree With best value of threshold (0.7) in RGB dataset	245057 pixels	96%
Bayesian Decision Tree With best value of threshold (0.1–0.9) in HSV dataset	245057 pixels	97%
Bayesian Decision Tree With best value of threshold (0.6) in YCbCr dataset	245057 pixels	93%

10 Conclusion

In this paper, human skin colour detection using BRDT was conducted in three experiments. Firstly, RGB dataset was collected from (UCI) machine learning repository. Secondly, HSV dataset was obtained from RGB dataset. Finally, YCbCr dataset was obtained from RGB. BRDT was used to improve skin detection without applying any image preprocessing steps. However, BRDT provides accuracy 98%, 97% and 97% in RGB dataset, HSV dataset and YCbCr dataset respectively. The high accuracy of skin detection with bright and dark illumination by using YCbCr dataset when the value of threshold is 0.6 that gives accuracy 93%.

References

1. Talukder, Z., Basak, A., Shoyaib, M.: Human skin detection. J. Comput. Sci. Technol. Graph. Vis. (2013)
2. Tan, W.R., Chan, C.S., Yogarajah, P.: A fusion approach for efficient human skin detection. IEEE Trans. Ind. Inform. **8**(1), 138–147 (2012)
3. Elgammal, A., Muang, C., Hu, D.: Skin detection - a short tutorial. In: Li, S.Z., Jain, A. (eds.) Encyclopedia of Biometrics. Springer, Boston (2009). https://doi.org/10.1007/978-0-387-73003-5_89
4. UCI machine learning repository. http://archive.ics.uci.edu/ml/datasets/Skin+Segmentation
5. Juang, C., Chiu, S.H., Shiu, S.J.: Fuzzy system learned through fuzzy clustering and support vector machine for human skin color segmentation. IEEE Trans. Syst. **37**, 1077–1087 (2007)
6. Bhatt, R., Dhall, A., Sharma, G., Chaudhury, S.: Efficient skin region segmentation using low complexity fuzzy decision tree model. In: 2009 Annual IEEE India Conference (2009)
7. Kolkur, S., Kalbande, D., Shimpi, P., Bapat, C., Jatakia, J.: Human skin detection using RGB, HSV and YCbCr color models. In: Iyer, B., Nalbalwar, S., Pawade, R. (eds.) ICCASP/ICMMD-2016. Advances in Intelligent Systems Research, vol. 137, pp. 324–332. Atlantis Press (2017)
8. Pujol, F., Pujol, M., Morenilla, A.J.: Face detection based on skin color segmentation using fuzzy entropy. Academic Editor, Raúl Alcaraz Martínez (2017)
9. Bush, I., Abiyev, R., Aitahand, M., Altıparmak, H.: Integrated artificial intelligence algorithm for skin detection. In: ITM Web of Conferences 16, 02004 (2018)
10. Shaika, K., Ganesan, P., Kalist, V., Sathish, B.S., Merlin, J.: Comparative study of skin – color detection and segmentation in HSV and YCbCr color space. Procedia Comput. Sci. **57**, 41–48 (2015)
11. Poorani, M., Prathiba, T., Ravindran, G.: Integrated feature extraction for image retrieval. Int. J. Comput. Sci. Mob. Comput. **2**(2), 28–35 (2013)
12. Zarit, B.D, Super, B.J, Quek, F.K.H.: Comparison of five color models in skin pixel classification. In: International Workshop on Recognition, Analysis, and Tracking of Faces and Gestures in Real-Time Systems, pp. 58–63 (1999)
13. Han, J., Kamber, M., Pei, J.: Data Mining: Concept and Techniques, 3rd edn. Morgan Kaufmann, San Francisco (2011)
14. Chaudhari, P., Rana, D.P., Mehta, R.G., Mistry, N.J., Raghuwanshi, M.M.: Discretization of temporal data: a survey. Int. J. Comput. Sci. Inf. Secur
15. Optimal Bin Number for Equal Frequency Discretization's in Supervised Learning

16. Han, J., Kamber, M.: Data Mining: Concepts and Techniques, 2nd edn. Morgan Kaufmann, San Francisco (2006)
17. Abbas, A.R., Juan, L., Mahdi, S.O.: A new version of the bayesian rough set based on bayesian confirmation measure. In: International Conference on Convergence Information Technology, pp. 284–289, China
18. Decision Tree – Data Mining Map. http://www.saedsayad.com/decision_tree

Offline Signature Recognition Using Centroids of Local Binary Vectors

Nada Najeel Kamal[1](✉) and Loay Edwar George[2](✉)

[1] Computer Sciences Department, University of Technology,
Iraq, Baghdad, Iraq
110136@uotechnology.edu.iq
[2] Department of Computer Science, College of Science,
University of Baghdad, Baghdad, Iraq
loayedwar57@yahoo.com

Abstract. The Handwritten Signature is a special sign used by humans and may contain letters, curves, or both. The main usage of handwritten signature is a proof of identification, especially when dealing with official documents and treatments. To recognize a signature means to identify the person who uses this sign. Signature recognition has many applications, such as: transactions and checks in banking systems, forensic caseworks, personal authentication and verification. This work proposes a new method to recognize handwritten signature in an offline manner. The centroid of two local binary vectors, the horizontal vector and the vertical vector are calculated. Three different tests are accomplished for this method. Soft evaluation test, Hard evaluation test, and a Combined test. The gained results from the three of these tests are encouraging. It reached for what is so called the Combined test to 94.8275% of success rate for 928 digital images of handwritten signatures with processing time for single sample reaches to 0.146 in milliseconds.

Keywords: Signature recognition · Local vector · Sub image
Centroid · Binary

1 Introduction

Identification of people is a very important aspect of today's life. It has many applications, especially where security aspects are needed. For many years, the required security is provided for people and governments through the technique of tokens, which was adopted and used in an extended manner. Examples of these used tokens may include: passwords, Identification cards, or PINs (Personal Identification Number), or badges. But this kind of technique has a major disadvantage of being forgotten or misused by individuals, and since biometrics are defined as the identification or verification of a person depending on his or her unique characteristics, day after day, biometrics are gaining more and more attention, and replacing the old fashion technique of using tokens [1–3].

© Springer Nature Switzerland AG 2018
S. O. Al-mamory et al. (Eds.): NTICT 2018, CCIS 938, pp. 255–268, 2018.
https://doi.org/10.1007/978-3-030-01653-1_16

There are two types of biometrics: physiological and behavioural. The first type defines an individual based on the measurement of his or her biological characteristics such as face, fingerprint, palm geometry, iris or retina. On the other hand, behavioural biometric defines a person based on his or her behavioural traits such as: speech patterns, gait, keyboard stroke, and handwritten signature [2, 4].

Handwritten signature is widely spread and accepted by the public as a means where people use it to identify themselves in daily life, particularly when dealing with official documents and treatments [4–6, 9]. This usage may include: handling banking accounts, checks, transactions, contracts, certificates, forensic casework, and employee attendance [4, 6–8].

Handwritten signature can be used for user recognition and verification. Recognition involves comparing a signature sample to set of N classes, while verification involves comparing a single sample to a single class [6].

Signature recognition and verification is divided into two types: online and offline. The first type concerns with processing of a signature directly, while it is being produced by the signer using tools such as a special pen and a tablet. While the second type processes the signature sample after storing it in a scanned digital image [1, 5–10].

Online also called "dynamic" is considered to be easier than the other type, because it collects information such as time, speed, pressure and stroke at the time of capturing the signature [9]. But dynamic recognition and verification has a major disadvantage, because it cannot be used in critical applications where the signer must present at the time of signing the signature [1].

On the other hand, offline recognition or verification, also called "static", is much cheaper [7], and there is no need for the signer to be present at the time of signing, but it is more difficult, because of the lack of the related time information [6, 7].

In general signature recognition is a challenging task because it depends on the psychological and physical situation of the signer, for example his age, health condition, emotions, … etc. It also depends of the body condition, since it is different when signing standing or sitting on a chair. The Environmental factors and conditions may affect the process of signing, these factors involve noise, lighting, and temperature. In addition to that, the signing tool and the signing surface has a major effect too, because it is quite different when using a pen, pencil or stylus on digital device, mobile or on paper [2, 11].

It is well known that the handwritten signature is widely used in the official treatments and documents, especially in the Iraqi governmental offices, and here the importance of this work comes from that if it is applied in these offices it can shorten the time and the routine of the work throughout finding, indexing, and archiving the many documents in an automated way, with less cost solution by recognizing a person's signature using the proposed static method.

This paper is concerned with producing a new proposed method to recognize the handwritten signature in an offline manner. The structure of this paper comes as the following: Sect. 2 contains a detailed review about many researches that tackle this aspect. Section 3 presents the preprocessing steps. Section 4 The feature extraction is given. Section 5 clarifies the classification used. In Sect. 6 the database is explored more in-depth. In Sect. 7 the Results are presented. Finally, Conclusions are in Sect. 8.

2   Related Work

In the research community there are many works and researches that have explored the subject of offline signature recognition and identification. Below, many of these researches are surveyed and reviewed in a detailed manner, what is written depends on the reported issues in these works.

Sigari et al. [1], preprocess the signature image by reducing the noise, normalizing image size to 200 pixels, rotating that image, and placing a virtual grid of distance 10 or 20 on it. Then, the Gabor wavelet operator is commuted for the points of that grid. The used classifier is weighted distance measure. The highest documented accuracy is to 100%.

Blanco-Gonzalo et al. [3] used Dynamic Time Wrapping (DTW) algorithm, where their DTW from previous work in [12] and [13] is tested on more mobile devices. The input of this algorithm is time, X and Y signals. 7040 signature images were used.

Patil and Patil in [4] preprocess the signature image by cropping it and scaling it to the size of 128×256. Then the global features are extracted from that image. These features include area, height, width, height to width ratio, and signature centroid. Another feature is sub dividing the image into four areas and calculating the summation of these sub features. The Euclidean Model is used as distance measure. 300 signature images are used for this research. 89% is the accuracy rate of the used method.

Neural Network (NN) is used in many researches. Daqrouq et al. [5] used methods which depend on Probabilistic Neural Network (PNN) for classification, and the entropy value of wavelet transform. Fixed image size of 80×250 is used. Accuracy of this method is 92%. Al-Saegh in [7] adopts the Weightless NN for recognition. In this research the image is preprocessed by: binarization, resizing to the size of 128×32 pixels, and thinning of that input image. Feature extraction is made by grids and rectangles of 4 or 64 or 256 cells only, for both of the horizontal and vertical directions. 1240 images were used to examine the proposed method. Results reached to 99.67%. The work of Karouni et al. [10] involves preprocessing where the coloured image is converted to a grey scaled one and binarization of that image, while features such as area, the centre of the coordinates, and skewness are extracted from the digital image. NN is used for classification. 100 signatures are used in this work. 93% is the success rate that is documented for this method. Połap and Woźniak [14] scale signature image to 40×15 pixels, then the inclination angle is adjusted. Then Chebyshev polynomials are created for the curves of the signature. Then a flexible NN is used for classification. 200 samples are used. The validation rate that is reported in this work reached to 93%.

Ubul et al. [6] preprocess the input signature image by: noise reduction, binarization, and normalization to the size of 384×96 pixels. A grid is placed on the image that has the size of 12×8 or 15×8 segments. The features are extracted based on the work of Abdala and Yousif [15]. Euclidean distance and K nearest neighbour are used as classifiers. 1000 images are tested. 93.53% is the attained accuracy of this work.

The Kudłacik and Porwik [8] method includes finding the signature centre of gravity, then crossed lines are generated along the signature. After that a fuzzy model is used for recognition. 1600 images are examined. Reported accuracy is 99.19%.

Porwik and Para [9] preprocess the image through the steps of binarization followed by cutting the edges and thinning. Feature extraction phase consists of the calculation of: width to height ratio, vertical and horizontal projections, centre of gravity and as basic issue Hough transform. To identify tested samples comparison coefficients are used. 800 genuine images are used. 94.60% is the documented success rate.

Fotak et al. [11] adopt graph theory for feature extraction, where graph is created for the signature. Comparison with database is used for classification. 400 tests are made. 94.25% is the accuracy of identification.

Nauom et al. [16] use complex moments to recognize the digital images of signatures. 11th orders of moment are tested with inclusion factor. Distance measure classifier is used. The number of used images is 144. These images come with rotation angles from 5° to 20° clockwise and counter clockwise. Success rates reaches to 99.31%.

Al-Ta'i, and Abdelhammed [17] preprocessed the image by binarizing it, and then applying thinning on that image. Moment algorithm is used for feature extraction and template creation. 100 signatures are used. The attained result reached to 90%.

In Chatterjee and Mukherjee [18] work, the preprocess steps are: converting the image into a binary form, cropping, image size normalization. The Euler Number is used as a feature that reflects the structure of the signature. 50 images are used. 80% is the success rate for this work.

3   Preparing Process

The process of preparing the digital image of the handwritten signature include three phases, they are: (1) converting the true coloured image into a grey scaled one, (2) binarize that grey image, and finally normalize its dimensions to be in a squared size image.

3.1   Obtaining the Grey Scaled Image

It is well known that Bitmap (BMP) digital image format is the most widely used format in image processing and pattern recognition issues, because it is a non-compressed format. Also, the colour depth variety of this format is familiar to the researchers. So, the BMP images are the used format in this proposed method, with 24 bits per pixels of colour depth of each of the digital image signature sample. This true coloured image of the three known bands: the red band, the green band and the blue band, is needed to be converted to a grey scaled image of one band, to facilitate the obtaining of binary images which is more suitable for the recognition steps. Equation (1) is used to get the grey image. Section 6 presents more details about the database of the signature images used in this work. See Fig. 1.

$$Grey(x, y) = Red(x, y) + Green(x, y) + Blue(x, y) \qquad (1)$$

where x and y are image pixels' location, $Grey()$ is the obtained grey image array, Red, $Green$, $Blue$ are the arrays of the original true coloured image of the signature.

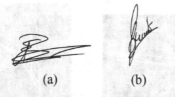

(a)   (b)

Fig. 1. Examples of hand written signatures. (a) is signature of traversal direction and (b) signature of vertical orientation

3.2 Digital Image Signature Binarization

Binary image is needed to be used in feature extraction phase. The binary image is acquired from the grey scaled image. The grey scaled image pixels values range from 0 to 255. A threshold of 127 is used to obtain the required binary image using the Eq. (2). Figure 2 shows examples of binary images, where the signature is considered the object that needs recognition and has a white colour, while the background of the image is in black.

$$Bin(x,y) = \begin{cases} 0 \text{ if } Gray(x,y) < Thr \\ 1 \text{ if } Gray(x,y) > Thr \end{cases} \tag{2}$$

$Bin()$ represents the binary image. x and y are image pixels Thr is the threshold value.

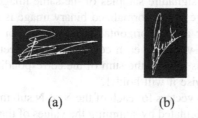

(a)   (b)

Fig. 2. Binary images. (a) and (b) are the binarized images of Fig. 1.

3.3 Normalization of Image Dimensions

Handwritten signatures come in different orientations, most of them are transversal, some are straight up, while others are diagonal, these different orientations made the segmented image of the handwritten signature come with different dimensions, i.e. the height of the image is greater than its width, or vice versa. Making the image dimensions equal is the goal of this phase. This equalization or image size normalization results in a squared sized image of equal height and width. This normalized image is needed in feature extraction phase. Equalization is made through padding the shortest dimension of image with zeros. Algorithm is available in [19]. See Fig. 3. This figure shows examples of image normalization.

Fig. 3. Normalized images of equal size in height and width

4   Features Extraction Using the Centroid of Local Binary Vectors

This phase is used to extract the features from the signature image, where these features are used to recognize and identify each single signature.

In this phase, the normalized binary image of the signature is divided into N × N local sub images. Note that the importance of dimension equalization appears here in an obvious manner, because the sub images must have an equal size of pixels, and if the normalization phase is not performed, then the size of these tiles or sub images will vary for the single image, which will affect the results of recognition and classification later.

These N × N sub images are not discrete, instead they are overlapped with each other with certain ratio. This overlapping is made to compensate for the little differences in shapes between signature samples of the same image class.

Then, every sub image in the normalized binary image is abstracted to two binary vectors, a vertical vector and a horizontal vector. A vector is represented in a one dimensionally array. The values in each cell in the local vector will be either 0 or 1. A cell of a vector contains 0 when the sum of the corresponding row pixels of the sub image equals to 0 otherwise it will hold 1.

After forming the two vectors for each of the N × N sub images, the centroid or the mean of each vector is calculated by summing the values of that vector and dividing the total by the number of cells of it. These centroids, i.e. the centroid of the vertical vector and the centroid of the horizontal vector of each of the sub images is considered to be the feature that extracted from the signature image. For example if an image is divided into 3 × 3 local sub images, this will result in (3 × 3) × 2 centroids or 18 feature values. The number of values in each vector equals to what is shown in Eq. (3)

$$M = N \times N \times 2 \tag{3}$$

Where is the obtained number of features from a single image, N represents the number of sub images, and 2 refers to the two centroids of vertical and horizontal vector, a single centroid for each vector.

A mean template vector and standard deviation template vector are created for each class of signature image that are divided by N × N of a certain number of sub images that are overlapped to certain ratios. Eight images from each class are used to create the template vectors. Figures 4 and 5 present feature extraction method.

Fig. 4. Local sub image of size 3 × 3. Two overlapped sub images are clarified in this image.

5 Classification Phase

The classification phase is used to specify the class of an input image sample. This classification is done by comparing the extracted features of the input image with the template values of all classes. The comparison is made throughout the distance measures. This research tests four formulas of Euclidean distance measures, they are shown in Eqs. (4–7). The desired class is determined by obtaining the minimum distance.

$$Dis_{i,j} = \sum_{s=1}^{M} \left(\frac{f_i(s) - \mu_j(s)}{\sigma_j(s)} \right)^2 \tag{4}$$

$$Dis_{i,j} = \sum_{s=1}^{M} \left| \frac{f_i(s) - \mu_j(s)}{\sigma_j(s)} \right| \tag{5}$$

$$Dis_{i,j} = \sum_{s=1}^{M} \left| f_i(s) - \mu_j(s) \right| \tag{6}$$

$$Dis_{i,j} = \sum_{s=1}^{M} \left(f_i(s) - \mu_j(s) \right)^2 \tag{7}$$

In these equations Dis represents the measured distance between the i_{th} feature vector $f()$, and the j_{th} template vector. s is the feature index number. The domain of the values of i and j depends on the number of used classes, i.e. 1–78. The μ refers to the mean template vector while σ denoted the standard deviation template vector.

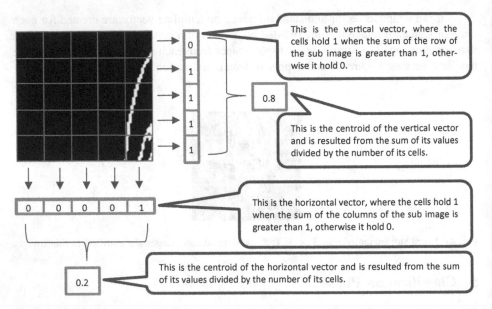

Fig. 5. The Feature extraction method, where the features are extracted as centroids of local binary vertical and horizontal vectors from a local sub image. Note: consider each square are in the enlarged sub image as image pixel.

6   The Used Database

The database that is used in this research is called NFI (Netherlands Forensic Institute). It is available online at URL [20]. In this web page, the link "evaluation set" is used, where the genuine signatures are tested. The zipped folder that contains the used samples comes under the name "6b_NFIgenuines.zip". The folder contains 940 in the format of Portable Network Graphics (PNG) of signature images, with size ranges from 1621 × 1100 to 440 × 235 pixels. Note that the researchers can provide you with the used database samples too.

Since the images of this database comes relatively in a large size, which consumes a long time for processing, and because BMP format is more simple and compatible with Microsoft Windows and has a well-known header information, that's why all of the images are downsized to 25% of its original size and converted to the BMP format. The application program which is used for this purpose is called "Ashampoo Photo Comander" version 11.0.3.

The names of the samples of this database comes as NFI – AAA-BBB-CC, where AAA represents the identification of the author, BBB represents the identification of the sample, while CC refers to the class number. This research renamed the names of the samples in each class, where the formula C(B) is used, where C is the class index, and the B is the class number. More details of this database are available in [21].

The used number of samples from this database is 928 not 940, because there are 12 samples are excluded. The excluded samples contain irregular shapes which are not compatible with other samples belonging to the referenced class. While some classes contain 5 samples only, and this number is not suitable for training or testing issues, such classes are excluded too. There are 78 used classes. The classes numbered from 1 to 71 contain 12 samples, while the remaining classes from 72 to 78 have 11 samples.

7 Recognition Test Results

The algorithm presented in this research is tested using the accuracy rate Eq. (8)

$$Accuracy = \frac{Number\ of\ correct\ Identifications}{Total\ Number\ of\ Samples} \times 100 \qquad (8)$$

Three types of tests are made. For every test the signature image is divided into a number of sub images ranges from $2 \times 2, 3 \times 3, 4 \times 4...$ to 100×100, and for every tested number of sub images the value of overlapping examined ranged from 0, 0.1, 0.2, ... 1. In each type of these tests, the four distance measures Eqs. (4–7) are examined and the acquired time is documented.

The first type of test called "Soft evaluation test". In this test all the samples which are used as training samples are tested. Where the first 8 samples of each class are used in this test. Table 1 shows that the attained accuracy is 100%.

Table 1. Soft evaluation test results

Total no. of samples	No. of samples correctly identified	Success rate percentage	Time in seconds	No. of local sub images	Ratio of overlapping	Distance measure equation no.
624	624	100	0.124	20×20	0.5	4
624	622	99.67949	1.465	88×88	0.4	7
624	613	98.23718	0.081	13×13	0.7	5
624	592	94.8718	0.117	20×20	0.3	6

The second type of test called "Hard evaluation test". The remaining samples from each class which are not used as training samples are included in the hard test. Since most of the classes consist of 12 samples, the images numbered 9, 10, 11 and 12 are the used samples in this test. Table 2 shows the results for this test. Depending on the numbering of this research the number of classes that contains 11 samples are 72–78.

Table 2. Hard evaluation test results

Total no. of samples	No. of samples correctly identified	Success rate percentage	Time in seconds	No.of local sub images	Ratio of overlapping	Distance measure equation no.
304	271	89.14474	0.139	24 × 24	1	7
304	260	85.52631	0.052	13 × 13	0.8	6
304	233	76.64474	0.051	13 × 13	0.7	5
304	220	72.36842	0.03	7 × 7	1	4

The last type of test is called "Combined evaluation test". In this test all of the samples of the classes were used. The samples included in both of the hard and soft test. That's why this test is called Combined test. Table 3 show the gained results from this test.

Table 3. Combined evaluation test results.

Total no. of samples	No. of samples correctly identified	Success rate percentage	Time in seconds	No. of local sub images	Ratio of overlapping	Distance measure equation no.
928	880	94.82758	0.146	26 × 26	0.5	7
928	846	91.1638	0.05	13 × 13	0.7	5
928	846	91.1638	0.091	19 × 19	0.8	6
928	826	89.00862	0.05	13 × 13	0.7	4

If the "Combined evaluation test" is considered as the ideal test, then depending on the result obtained from it, the accuracy reached to 94.82758%, where 880 samples are correctly recognized from the total of 928 signature images. See Table 3. The time reached to 0.146 in seconds. The accuracy rate is gained when the Eq. (7) of distance measure is used, with number of sub images 26 × 26 with rate of overlapping of (0.5).

Note that what is shown in Tables 1, 2 and 3 the highest attained result for each distance measure. The following code list gives more clarification about the used tests and how the results are collected and interpreted.

```
Begin
For each number N of local sub images Loop {N from 2×2 to
100×100}
  For each ratio of overlapping R Loop {R=0, 0.1, 0.2… 1}
    Total=0; TruePositive=0; Percentage=0
    For every class number C Loop {C=1 to 78}
      NumberofPassedSample=0: EnteredSamples=0;
      For every sample Q in the class Loop
        EnteredSamples= EnteredSamples+1;
        Convert image to gray;
        Convert image to binary;
        Normalize image size;
        Get the centroid of vertical and horizontal vector
        of the sub image;
        Compare the extracted features with mean template
        and standard deviation template;
        Determine the closest class number;
        NumberofPassedSample= NumberofPassedSample+1;
      End Loop {of Q}
      Total= Total+ EnteredSamples;
      TruePositive= TruePositive+ NumberofPassedSample;
      Percentage= (NumberofPassedSample/ EnteredSam-
      ples)*100;
      Document consumed time;
    End Loop {of C}
  End loop {of R}
End Loop {of N}
End
```

The details of the best attained result are shown in Table 4. The system is tested on 32- bit, Windows 7 operating system. Intel Core i3 CPU, of 6 GB of memory. The software is built using visual basic language.

Table 4. Details of result of combined test for each class when the number of sub images is 26 × 26, with rate of overlapping of 0.5, and the Eq. 7 is used.

Number of class		Total number of class samples	Correctly identified samples	Accuracy percentage	Number of class		Total number of class samples	Correctly identified samples	Accuracy percentage
NFI class no.	C				NFI class no.	C			
01	1	12	11	91.66666	59	41	12	11	91.66666
02	2	12	12	100	62	42	12	11	91.66666
03	3	12	10	83.33334	63	43	12	11	91.66666
04	4	12	12	100	64	44	12	11	91.66666
06	5	12	12	100	66	45	12	12	100
07	6	12	12	100	67	46	12	11	91.66666
08	7	12	12	100	68	47	12	11	91.66666
10	8	12	12	100	69	48	12	10	83.33334
11	9	12	12	100	70	49	12	12	100
12	10	12	12	100	72	50	12	11	91.66666
14	11	12	11	91.66666	73	51	12	12	100
15	12	12	12	100	74	52	12	12	100
16	13	12	12	100	75	53	12	11	91.66666
17	14	12	12	100	77	54	12	10	83.33334
21	15	12	12	100	79	55	12	12	100
22	16	12	12	100	80	56	12	10	83.33334
23	17	12	10	83.33334	83	57	12	11	91.66666
24	18	12	12	100	85	58	12	12	100
26	19	12	12	100	86	59	12	12	100
27	20	12	12	100	88	60	12	12	100
28	21	12	10	83.33334	89	61	12	12	100
29	22	12	10	83.33334	90	62	12	12	100
31	23	12	11	91.66666	91	63	12	12	100
33	24	12	12	100	92	64	12	12	100
35	25	12	12	100	93	65	12	12	100
37	26	12	11	91.66666	94	66	12	11	91.66666
41	27	12	11	91.66666	96	67	12	10	83.33334
42	28	12	11	91.66666	97	68	12	12	100
43	29	12	9	75	98	69	12	12	100
44	30	12	12	100	99	70	12	11	91.66666
45	31	12	11	91.66666	00	71	12	12	100
46	32	12	12	100	09	72	11	11	100
47	33	12	11	91.66666	19	73	11	10	90.90909
49	34	12	11	91.66666	20	74	11	11	100
51	35	12	10	83.33334	30	75	11	11	100
53	36	12	10	83.33334	39	76	11	11	100
54	37	12	11	91.66666	71	77	11	11	100
55	38	11	11	100	84	78	11	10	90.90909
56	39	12	12	100	Total		928	880	94.8275%
58	40	12	10	83.33334					

8 Conclusions and Future Work

This research presents an excellent method to recognize and identify handwritten signature images. The used algorithm extracts the feature from the digital image by dividing the image into local sub images, then abstracting each of these tiles into two binary vectors. The mean of each vertical and horizontal vector is used as a basic feature extracted from that sub image.

The number of tested images is 928; they are examined using four statistical distance measurements. Three types of experimental tests are made: soft, hard, and combined. The accuracy gained based on the combined test is 94.82758%.

The presented method is shape, style and image size independent, and therefore it can be used for signature verification and signature recognition, or it may be used for documents indexing, archiving, or retrieving. Further development for this method may involve extracting other statistical parameters from the local horizontal and vertical vectors such as standard deviation and variance. Also, it is legible to say that using faster hardware for applying or testing this system may shorten the processing time of identification.

References

1. Sigari, M.H., Pourshahabi, M.R., Pourreza, H.R.: Offline handwritten signature identification and verification using multi-resolution gabor wavelet. Int. J. Biom. Bioinform. **5**(4), 234–248 (2011)
2. Impedovo, D., Pirlo, G.: Automatic signature verification: the state of the art. IEEE Trans. Syst. Man Cybernet. Part C Appl. Rev. **38**(5), 609–635 (2008)
3. Blanco-Gonzalo, R., Miguel-Hurtado, O., Mendaza-Ormaza, A., Sanchez-Reillo, R.: Handwritten signature recognition in mobile scenarios: performance evaluation. In: IEEE International Carnahan Conference on Security Technology, pp. 174–197. IEEE Press, Boston (2012). https://doi.org/10.1109/ccst.2012.6393554
4. Patil, P., Patil, A.: Offline signature recognition using global features. Int. J. Emerg. Technol. Adv. Eng. **3**(1), 408–411 (2013)
5. Daqrouq, K., Sweidan, H., Balamesh, A., Ajour, M.N.: Off-line handwritten signature recognition by wavelet entropy and neural network. Enropy 1–2031 (2017). https://doi.org/10.3390/e19060252
6. Ubul, K., Adler, A., Abliz, G., Yasheng, M., Hamdulla, A.: Off-line Uyghur signature recognition based on modified grid information features. In: 11th International Conference Information Science, Signal Processing and their Applications, pp. 1056–1061 (2012)
7. Al-Saegh, A.: Off-line signature recognition using weightless neural network and feature extraction. Iraq J. Electr. Electron. Eng. **11**(1), 124–131 (2015)
8. Kudłacik, P., Porwik, P.: A new approach to signature recognition using the fuzzy method. Pattern Anal. Appl. **17**(3), 451–463 (2014). https://doi.org/10.1007/s10044-012-0283-9
9. Porwik, P., Para, T.: Some handwritten signature parameters in biometric recognition process. In: IEEE Conference Information Technology Interfaces, pp. 185–190. IEEE Press, Cavtat (2007). https://doi.org/10.1109/iti.2007.4283767
10. Karouni, A., Daya, B., Bahlak, S.: Offliner signature recognition using neural networks approach. Procedia Comput. Sci. **3**, 155–161 (2011). https://doi.org/10.1016/j.procs.2010.12.027

11. Fotak, T., Bača, M., Koruga, P.: Handwritten signature identification using basic concepts of graph theory. Wseas Trans. Sig. Process. **7**(4), 145–157 (2011)
12. Pascual-Gaspar, J.M., Cardeñoso-Payo, V., Vivaracho-Pascual, C.E.: Practical on-line signature verification. In: Tistarelli, M., Nixon, M.S. (eds.) ICB 2009. LNCS, vol. 5558, pp. 1180–1189. Springer, Heidelberg (2009). https://doi.org/10.1007/978-3-642-01793-3_119
13. Mendaza-Ormaza, A., Miguel-Hurtado, O., Blanco-Gonzalo, R., Diez-Jimeno, F.J.: Analysis of handwritten signature performances using mobile devices. In: IEEE Translation of International Carnahan Conference on Security Technology (2011)
14. Połap, D., Woźniak, M.: Flexible neural network architecture for handwritten signatures recognition. Int. J. Electron. Telecommun. **62**(2), 197–202 (2016). https://doi.org/10.1515/eletel-2016-0027
15. Abdala, M.A., Yousif, N.A.: Offline signature recognition and verification based on artificial neural network. Eng. Technol. J. **27**(7), 1–9 (2009)
16. Nauom, R.S., Jorj, L.A., Musa, A.K.: Signature recognition by using complex-moments characterisics. Iraqi J. Sci. **43D**(3), 11–23 (2002)
17. Al-Ta'i, Z.T., Abdelhammed, O.Y.: Off-line signature identification using moment algorithm. Diyala J. Pure Sci. **8**(2), 1–18 (2012)
18. Chatterjee, S., Mukherjee, J.: Handwritten signature recognition system using euler number. Int. J. Adv. Inf. Sci. Technol. **25**(25), 16–19 (2014)
19. Kamal, N.N.: Utilization of edge information in handwritten numeral recognition. Iraqi J. Sci. **57**(2A), 984–994 (2016)
20. SigComp2009. http://www.iaprtc11.org/mediawiki/index.php/ICDAR_2009_Signature_Verification_Competition_(SigComp2009)
21. Blankers, V.L., van den Heuvel, C.E., Franke, K.Y., Vuurpijl, L.G.: The ICDAR 2009 signature verification competition. In: 10th IEEE International Conference on Document Analysis and Recognition. IEEE Press, Barcelona (2009). https://doi.org/10.1109/icdar.2009.216

Automatically Recognizing Emotions in Text Using Prediction by Partial Matching (PPM) Text Compression Method

Amer Almahdawi[1](✉) ⓘ and William John Teahan[2](✉)

[1] College of Science for Women, Baghdad University, Baghdad, Iraq
amer.almahdawi@gmail.com
[2] School of Computer Science, Bangor University, North Wales, UK
w.j.teahan@bangor.ac.uk

Abstract. In this paper, we investigate the automatic recognition of emotion in text. We perform experiments with a new method of classification based on the PPM character-based text compression scheme. These experiments involve both coarse-grained classification (whether a text is emotional or not) and also fine-grained classification such as recognising Ekman's six basic emotions (*Anger, Disgust, Fear, Happiness, Sadness, Surprise*). Experimental results with three datasets show that the new method significantly outperforms the traditional word-based text classification methods. The results show that the PPM compression based classification method is able to distinguish between emotional and nonemotional text with high accuracy, between texts involving *Happiness* and *Sadness* emotions (with 80% accuracy for Aman's dataset and 76.7% for Alm's datasets) and texts involving Ekman's six basic emotions for the LiveJournal dataset (87.8% accuracy). Results also show that the method outperforms traditional feature-based classifiers such as Naïve Bayes and SMO in most cases in terms of accuracy, precision, recall and F-measure.

1 Introduction

Affective computing (or artificial emotional intelligence) is the study of computer systems that are able to recognise and respond to human affects (their feelings and emotions). Recognising a person's emotional state is possible using such cues as their facial expressions, their voice, the language they use or their behaviour. Written text such as email, texting, blogs and tweets now makes up a significant amount of the communication between people because of the growth of social media. Therefore, being able to recognise the emotional state of the person or persons producing the text would be very beneficial in many situations. For example, recognising certain types of emotion might help to predict when someone might commit a crime or a terrorist act [23] or provide early risk detection for the Internet, particularly in health and safety areas such as detecting early signs of depression, anorexia or suicidal inclinations [18]. Another area where emotion recognition is useful is in helping to build more affective interfaces where identifying the emotion of the user can allow the computer to respond more effectively.

© Springer Nature Switzerland AG 2018
S. O. Al-mamory et al. (Eds.): NTICT 2018, CCIS 938, pp. 269–283, 2018.
https://doi.org/10.1007/978-3-030-01653-1_17

Emotion recognition in text is a specific type of text categorisation. Shaheen et al. [19] state that there are two types of emotion classification: those that are coarse grained and those that are fine grained. Coarse grained classification tries to identify positive and negative emotions in the text as occurs for sentiment analysis. Fine grained classification on the other hand tries to identify more than just the two positive and negative categories by identifing more specific emotions (such as Happiness and Sadness). Ekman has stated that there are six basic emotions—Anger, Disgust, Fear, Happiness, Sadness, and Surprise—as these emotions are common to all cultures [7, 8]. Evidence for this was obtained by examining brief facial expressions that occur when a person is trying to conceal an emotion either deliberately or unconsciously.

The purpose of this paper is to determine the effectiveness of one possible method for recognising emotions in text using text compression. We are interested in both coarse grained emotion recognition (such as whether the text is emotional or not) and fine-grained emotion recognition such as distinguishing between Happiness and Sadness emotions or distinguishing texts according to Ekman's six basic emotions. Text compression can be used to classify texts by emotion using a two stage supervised learning process: the first stage builds models by training on texts that are representative of each type of emotion being classified; and the second stage uses the training models to compress each testing text, and then assign the class using the label associated with the model that compresses the testing text best.

This paper is organised as follows. The next section discusses related work. The PPM-based method for classifying texts is then discussed followed by a description of the datasets used in our experiments along with the experimental results. The paper completes with the conclusion and future work in the final section.

2 Related Work

As mentioned before, many researchers have investigated coarse grained classification (sentiment analysis and opinion mining) although less so in the area of emotion recognition. Although we investigate course-grained classification in the sense of distinguishing between emotional and non-emotional texts, this is different from sentiment and opinion. This paper will also explore fine grained classification by investigating the automatic recognition of specific emotions.

Alm et al. [1] conducted early work on recognising in text the emotions defined by Ekman by using 22 children fairy tales as the dataset. Aman et al. [3, 4] built a corpus from web blogs and used the Naïve Bayes [13] and support vector machine (SMO) [5, 9] classifiers to classify test data also according to Ekman's basic emotions. Keshtar [11] used similar techniques combined with decision trees and experimented with a corpus built from Live Journal web blogs. These web blogs contained further information about the mood or moods of the writer of the blog. 132 moods were used in total such as excited and astonished. Various further studies have also experimented with the same LiveJournal dataset [13, 14].

Aman et al. [3, 4] built another corpus from web blogs that were classified by Ekman's six basic emotions and experimented with the Naïve Bayes and the support vector machine (SMO) classifiers. Chaffer and Inkpen [5] experimented with various datasets including Alm's dataset, Aman's dataset, a dataset consisting of news head-lines [20], and Neviarouskaya's dataset [16, 17]. The following four classifiers were investigated: ZeroR, Naïve Bayes, J48 and SMO [5]. Ghazi et al. [9] used a multilevel hierarchy combined with SMO as the classifier to classify Ekman's six basic emotions for both Alm's and Aman's datasets.

The previous research detailed above has used feature-based approaches (i.e. using words) for emotion recognition by applying standard machine learning classifiers such as Naïve Bayes or support vector machine SMO. An alternative information theoretic approach based on text compression has been found to be effective for many text categorisation tasks [22] but this has yet to be fully applied to the problem of emotion recognition. The unique aspect of this approach is that by processing all the characters in the text using text compression algorithms, it avoids issues to do with feature extraction and the need to process texts into words which can be problematical. However, the success of this approach on the specific problem of emotion recognition has yet to be investigated fully in the literature. Our preliminary experimental findings on this research were first published in [2]. The primary purpose of this paper is to report further experimental findings with this new approach.

3 PPM-Based Text Categorisation

Prediction of Partial Matching (PPM) is an adaptive lossless text compression method first published in 1984 [6] that processes characters in the text in a sequential manner. A variable order Markov-based model is updated dynamically as the text is processed with both the encoder an decoder maintaining the same model at each stage of the encoding and decoding processes. A finite context is used to predict the upcoming character in the text. Standard PPM will use a fixed maximum context to try to make its initial prediction. (This defines the "order" of the model). Text compression experi-ments with English and other natural language texts have shown that a fixed maximum context length of 5 (i.e. an order 5 model) usually works best. The method essentially estimates probabilities for the upcoming character.

The model uses an "escape" mechanism that smoothest the probability estimates by backing-off to a shorter context when novel characters are encountered (i.e. those with zero probabilities). This backing-off process may need to be undertaken multiple times until a context is found where the character can be predicted. For characters that have not been seen anywhere previously in the text, a default order −1 context is used where every character is predicted with equal probability. Various escape methods (such as methods A, B, C and D) have been devised over the years to define how the escape probability is estimated. (These are described in the literature as variants PPMA, PPMB and so on).

The PPMC variant was developed by Moffat [15] and has become the benchmark version. The probability estimates for this method is based on using the number of characters that have occurred before, called the number of types:

$$e(X) = \frac{t(X)}{f(X) + t(X)} \quad and \quad p(x_i/X) = \frac{C(X_i/X)}{n(X) + t(X)}$$

where $e(X)$ represents the probability of the escape symbol for context X, $p(x_i|X)$ denotes the probability for character x_i given context X, $c(x_i|X)$ is the number of times the context X was followed by the character x_i, $f(X)$ is the total number of times that the context X has occurred and $t(X)$ denotes the total number of types in that context.

The PPMD variant was first developed by Howard in 1993 [10]. In most cases, experiments show that the PPMD performs better than the other variants. This variant is similar to the PPMC variant with the exception that each count is incremented by a 1/2:

$$e(X) = \frac{t(X)}{2f(X)} \quad and \quad p(x_i/X) = \frac{C(X_i/X) - 1}{2f(X)} \tag{1}$$

Performance of the PPM models is improved using two mechanisms. The first is called 'full exclusions' and involves excluding counts from lower order calculations during escaping for symbols already predicted by a higher order (since they would have been encoded already so can be excluded). The second is called 'update exclusions' which involves updating counts in a context for a symbol only if it has not already been seen at a higher order. Both mechanisms have been found to improve compression by a few percent in most experiments. The PPM model can be represented by the following formula:

$$H_M(T) = \sum_1^n - \log_2 p\left(x_i/x_{i-m,...x_{i-1}}\right)$$

where HM is the compression code length given model M of order m for the probability distribution for the characters xi over the text sequence T = x1, x2 ...xn of length n. Each character will be predicted based on the prior context xi−m, ..., xi−1 of length m.

In order to better illustrate some important aspects of how PPM works using a relevant example, Table 1 shows a dump of the PPM model after processing the blog in Table 2 that starts with the text "the trip was fantastic...". (Note that for this example, one further unique end of text character has also been added at the end of the blog text). The model has been stored in a trie data structure with a maximum depth of 6 (since an order 5 model is being used) which stores all the suffixes up to length 6 that are contained in the text. However, the table only shows the root node (top left) along with nodes associated with suffixes starting with the letter 'a' (all the rest of the table). In reality, the table would be much bigger if all the suffixes (that do not start with the letter 'a') were included as well.

Each set of columns in the table that are separated by the double vertical bar contains information concerning the nodes in the trie; the nodes are arranged using a preorder traversal. There are separate columns for the node counts, the first being for PPMC without update exclusions (which we have labelled as PPMC0 as normally PPMC would perform update exclusions), while the second set of counts are for standard PPMD with update exclusions. The column labelled 'Path to trie node' shows the path down the trie to each node and the column labelled D indicates the depth of each node in the trie. The space character is indicated in the table by the ' ' character in the trie path. The model does not store counts when the context becomes unique; instead it stores a pointer back into the input text to allow for future updates. This is indicated in the table with a '–' in the counts columns. The counts in these cases are set to 1 for both the PPMC0 and PPMD models.

Table 1 shows a clear difference in the counts that are stored for PPMC0 and PPMD with the PPMC0 counts accurately reflecting the raw counts in the source text. For example, referring to the top two lines of the table across the second, sixth and tenth columns: the number of times any character occurs including the final end of text character is 491 times (at depth 0 of the trie); 'and' occurs 8 times; 'abl' occurs 2 times; the letter 'a' occurs 44 times; 'and' followed by a space character occurs 8 times (i.e. 'and' is always followed by a space); and 'able' occurs 2 times (i.e. 'abl' is also always followed by the letter 'e'). On the third line, the suffix 'and t' becomes unique, so is indicated by a '–' in the counts columns. In contrast, the PPMD counts are clearly different—many of the PPMD counts are equal to 3 which is twice the equivalent PPMC0 count (2) minus 1, as per Eq. 1. For many nodes the counts are much less than the PPMC0 counts due to the way the update exclusion mechanism works (which, as stated, only updates counts for the longest context if the symbol being predicted in that context has not already been seen).

The counts stored in this trie data structure are then used to estimate the probabilities as each character in the text being encoded is processed sequentially. For example, if the text being processed is "fantastic!", then the probabilities for each character using the static PPMD model stored in Table 1 would be calculated as follows. For the first character 'f' there is no prior context, so an order 0 context is used to estimate the probability, with a probability of $\frac{11}{362}$. For the second character 'a', an order 1 context can be used, with a probability of $\frac{3}{12}$. The order increases with subsequent characters up to the maximum order of 5. The contexts for these characters, however, have not been included in Table 1 due to space limitations (as stated, only the contexts beginning with 'a' have been shown). We encounter a context which is shown in the table when we encode the second 't' character (after the order 5 context 'antas'). Here the probability is estimated as $\frac{3}{4}$ where the numerator 3 is the PPMD count listed in the third column on the second to last line of the table. When the '!' character is encountered at the end, this is not found in the order 5 context 'astic' (see the suffixes for 'astic' in column 4 of the table) so the model has to escape from order 5 down to order 1, and again down to order 0 where the '!' character is found. Thus the probability that is encoded is $\frac{2}{4} \times \frac{6}{12} \times \frac{1}{256}$.

Text categorisation using PPM is performed by training N different models M1, M2,...MN where N is the number of classes and the training text used to train each

Table 1. Dump of PPM trie for the root node and paths starting with the letter 'a' after the sample blog starting with the text "the trip was fantastic..." in Table 2 has been processed. The space character is represented by ' '.

D	PPMC⁰ counts	PPMD counts	Path to trie node	D	PPMC⁰ counts	PPMD counts	Path to trie node	D	PPMC⁰ counts	PPMD counts	Path to trie node
0	491	58		3	8	1	and	3	2	1	abl
1	44	25	a	4	8	1	and_	4	2	1	able
2	10	3	as	5	–	–	and t	5	2	1	able_
3	8	1	as_	5	2	1	and a	6	–	–	able _b
4	2	1	as f	6	2	3	and ad	6	–	–	able _a
5	2	1	as fa	5	–	–	and j	2	2	3	ac
6	2	3	as fan	5	–	–	and c	3	–	–	aca
4	–	–	as o	5	–	–	and n	3	–	–	ace
4	–	–	as c	5	2	1	and i	2	2	3	a_
4	–	–	as g	6	2	3	and i	3	2	3	a _w
4	2	1	as t	2	5	7	al	4	2	3	a _wa
5	2	1	as th	3	–	–	aL	5	2	3	a _was
6	2	3	as the	3	–	–	ale	6	2	3	a was-
4	–	–	as s	3	2	3	all	2	–	–	ap
3	2	1	ast	4	–	–	ally	2	2	3	ar
4	2	1	asti	4	–	–	alls	3	–	–	arl
5	2	1	astic	3	–	–	alo	3	–	–	ara
6	–	–	astic,	2	4	5	ad	2	2	3	at
6	–	–	astica	3	2	1	ado	3	2	3	at_
2	11	5	an	4	2	1	ador	4	2	3	at w
3	3	3	ant	5	2	1	adora	5	–	–	at wa
4	2	1	anta	6	2	3	adorab	5	–	–	at wi
5	2	1	antas	3	–	–	ade	2	–	–	ag
6	2	3	antast	3	–	–	ad_	2	–	–	au
4	–	–	ant	2	2	1	ab	2	–	–	av

model is representative of the class being modelled. The main idea is to guess the correct class of the text T using the following formula:

$$\hat{\theta}(T) = \arg\min_i H_{Mi}(T)$$

for each class i. Essentially, one constructs a model for each class, and the text is compressed using each model with the class being chosen from the model that compresses the text best.

4 Datasets for Emotion Recognition

Three datasets have been used for the experiments described in this paper. A full description of these datasets is provided below.

Some samples from each of the datasets have also been provided in Table 2. They show the diverse nature of the texts included in the datasets and that for the blogs especially, the text contains many non-standard features including spelling mistakes, grammatical errors and colloquialisms potentially making the classification task more difficult.

LiveJournal Dataset The LiveJournal dataset is a large dataset composed of 815,494 web blog posts (whose total size is 1.6 GB in the original XML format). LiveJournal is a free weblog service available at http://www.LiveJournal.com/ used by millions of users. It is classified into 132 moods such as happy, cheerful and sad where the author of the blog has chosen to describe their mood while writing his/her post [14]. One issue with this dataset is that the consistency of the moods found in the dataset is highly variable because they are individually assigned. On the other hand, it helps us to access the writer's emotion directly without using an external annotator.

Table 2. Samples from the datasets used in our experimental evaluation.

Dataset	Emotion	Sample
LiveJournal	*Anger*	.... i have nothing positive to say right now. at all
LiveJournal	*Fear*	i feel like the world is talking behind my back.... i feel like the person who is the but of the joke but doesnt know it....
LiveJournal	*Happiness*	Well kids, I had an awesome birthday thanks to you. = D Just wanted to so thank you for coming and thanks for the gifts and junk. =) I have many pictures and I will post them later. 'hearts;
Aman	*Disgust*	I think the most important thing I can say about this city, is that all of the rumors: That the city's dirty, the people are rude, the language barrier is insurmountable, the metro is incomprehensible, you things will be stolen, that they hate Americans
Aman	*Happiness*	the trip was fantastic, plimoth was old and there were real live pilgrims everywhere, salem was cute and adorable but very touristy, ithaca was gorgeous and adorable and joel was there so the whole thing made me very happy and clearly that was the highlight of the trip, niagara was fantastically beautiful but boring once you've seen the falls, buffalo was sketchy but had great wings and now im home, and i think it's the most wonderful place ive ever been and i dont ever want to leave!
Alm	*Anger-Disgust*	Then he was very angry, and went without his supper to bed; but when he laid his head on the pillow, the pin ran into his cheek: at this he became quite furious, and, jumping up, would have run out of the house; but when he came to the door, the millstone fell down on his head, and killed him on the spot
Alm	*Sadness*	And now the sister wept over her poor bewitched brother, and the little roe wept also, and sat sorrowfully near to her
Alm	*Surprise*	But–seated upon the stump, she was startled to find an elegantly dressed gentleman reading a newspaper

Aman's Dataset This dataset is composed of emotion-rich sentences taken from web blogs. The author of this dataset searched for web blogs that contained various seed words representing the emotion. This dataset consists of 1290 emotional sentences of six types of emotions (179 Anger sentences, 172 Disgust sentences, 115 Fear sentences, 536 Happiness sentences, 173 Sadness sentences, and 115 Surprise sentences) based on Ekman's definition of emotions and also comprises 2800 non-emotional sentences in total. A collection of seed words was used for each emotion; for instance, the Happiness emotion encompasses the following seed words ('awesome', 'happy', 'amused', 'fantastic', 'excited', 'pleased', 'cheerful', 'love', 'great', 'amazing'). It also uses a neutral label for sentences that do not contain emotions. Four annotators were used to manually label sentences in the resulting dataset [3, 4].

Fairy tales (Alm's) Dataset This dataset consists of annotated sentences taken from fairy tales. The sentences in this dataset are labelled with five types of emotions (Anger-Disgust, Fear, Happiness, Sadness, and Surprise) based on Ekman's definition of emotion. Since Anger and Disgust emotions are similar, Alm decided to merge them into one class [1].

5 Experimental Results

Our experimental results are presented in this section. The purpose of these experiments was to evaluate the effectiveness of the proposed method. The PPM compression codelength estimates were obtained using the Text Mining Toolkit developed by Teahan and described by Mahoui et al. [12]. This toolkit allows both dynamic and static PPM models to be constructed from training text. Static PPM models are not updated subsequently on the testing test once the models have been trained unlike standard dynamic PPM models. Previous text categorisation experiments [22] have shown that these models compared to dynamic models are more efficient in terms of speed and memory utilisation as well as being very effective in terms of classification performance in most cases. Static order 5 PPM models with escape method D were created using training texts for each class (as these models usually lead to the best compression for English text). These models were then used to classify the separate testing data using the process described in Sect. 3.

We pre-processed the various datasets prior to each of the experiments described below in the following way. Aman and Alm's datasets are directly labelled by Ekman's basic emotions so therefore no further pre-processing was required. Aman's dataset provides an extra label for blogs that do not contain emotions. In Alm's dataset, each sentence is directly classified with an emotion label according to Ekman's system. However, further pre-processing was required for the LiveJournal dataset as follows: the XML tags, punctation and URLs were removed first; the blogs labelled in the same class were then extracted and concatenated together to form six separate text files for Ekman's six classes [2]. Aman and Szpakowicz [3] used seed words to collect blogs from the web. These seed words can be used to map the 132 moods to the six Ekman emotions using synonyms (such as using the words awesome and fantastic as synonyms for the Happiness emotion). The text files were then further split into ten partitions for the ten-fold cross-validation process.

The experiments we conducted using these texts are described in the next three sub-sections. We first describe coarse-grained classification experiments for distinguishing emotional versus non-emotional texts. Next we describe finegrained binary classification experiments to distinguish texts involving just the Happiness and Sadness emotions. Finally, we describe fine-grained classification experiments to distinguish texts involving Ekman's six basic emotions.

5.1 Experiments with *Emotional* Versus *Non*-emotional Sentences for Aman's Dataset

In this experiment, Aman's dataset has been used for the training and testing data. The purpose of this experiment was to evaluate the effectiveness of the PPM classifier at distinguishing between emotional and non-emotional content. Based on the available training data in Aman's dataset, we used text to train both the emotional and non-emotional PPM models. Two text files were extracted from Aman's dataset. One of these files contained 1290 blogs deemed to be emotional, while the other file contained 2770 blogs deemed to be nonemotional. These text files were used to evaluate the PPM classifier using a ten-fold cross-validation process with 9/10 of the text being used to train two static order 5 PPMD models which were then used to predict the appropriate class on the remaining test data.

The ZeroR, J48, Naïve Bayes and SMO classifiers implemented in Weka were applied to the same dataset. For all classifiers, the StringToWordVector filter has been used with the NGramTokenizer to select NGrams as features to compare with PPM since the latter implicitly works with n-grams. InfoGainAttributeEval has been used as the attribute evaluator using the Ranker search method on emotional versus non-emotional classification with the best features. The results that were obtained are shown in Table 3.

As shown in Table 3, the highest accuracy was achieved by the Naïve Bayes and SMO classifiers with 68.8%, while the PPM classifier achieved the lowest accuracy among these classifiers with 65.6%. However, in terms of precision, recall and F-measure, the PPM classifier clearly outperforms the feature-based classifiers that we experimented with such as ZeroR, Naïve Bayes, J48, and SMO. The precision of the feature-based classifiers were as follows: ZeroR 0.50, Naïve Bayes 0.50, J48 0.50, and SMO 0.50. The PPM classifier achieved a significantly higher precision of 0.62. In addition, PPM achieved the best results for recall and F-measure compared to other classifiers with 0.63 recall and 0.63 F-measure. While ZeroR and J48 achieved 0.50 for both recall and F-measure, Naïve Bayes and SMO achieved recall results of 0.51 and F-measure of 0.50. All measures were computed as macro-averages of precision, recall, and F-measure for the emotional and non-emotional classes for the ten folds used during the ten-fold cross-validation evaluation process.

Table 3. Classification results on emotional versus non-emotional sentences for different classifiers on Aman's dataset.

Classifier	Accuracy	Precision	Recall	F-measure
PPM	65.6%	0.62	0.63	0.63
ZeroR	68.3%	0.50	0.50	0.50
Naïve Bayes	68.8%	0.50	0.51	0.50
J48	68.3%	0.50	0.50	0.50
SMO	68.8%	0.50	0.51	0.50

5.2   Binary Classification: *Happiness* Versus *Sadness*

The second set of experiments investigated the binary classification problem of distinguishing between texts classed by the two Ekman emotions Happiness and Sadness. These emotions are the only pair of Ekman's emotions that are antonyms of each other and therefore they should be easier to distinguish.

The PPM method was first applied to the data extracted from the LiveJournal data for just the two classes. The two text files used for our experiments contained 503 blogs in the Happiness class, and 1258 blogs in the Sadness class. The results of the experiment are presented in Table 4 where we compared PPM results with other classifiers such as Naive bayes, ZeroR, J48, and SMO on the same tested blogs.

In comparison, Mihalcea and Liu also used LiveJournal blogs to classify only the Happiness and Sadness blogs by using a Naïve Bayes classifier and their method achieved 79.13% accuracy [13] but Mihalcea used five-fold cross validation.

Our experimental results in distinguishing Happiness versus Sadness are shown in Table 4. They show that PPM outperforms the other classifiers in terms of accuracy, recall and F-measure, although Naïve Bayes and SMO have better precision.

Table 4. Classification results on Happiness versus Sadness sentences for different classifiers on LiveJournal's dataset.

Classifier	Accuracy	Precision	Recall	F-measure
PPM	76.0%	0.81	0.59	0.68
ZeroR	71.1%	0.36	0.50	0.42
Naïve Bayes	71.3%	0.86	0.50	0.63
J48	71.1%	0.36	0.50	0.42
SMO	72.2%	0.86	0.51	0.64

The next experiments investigated the Happiness versus Sadness binary classification for Aman's and Alm's datasets. For Aman's dataset, text was extracted for the two different classes. The number of happiness sentences to be tested was 530, and the number of sadness sentences was 170 sentence. On the other hand, Alm's dataset consisted of 440 sentences for the happiness emotion, while there were 260 sentences for the sadness emotion. These texts were used to train PPM models and classify test data separately. PPM was used to produce models of each text. Ten-fold cross validation was used to evaluate the classification of test data according to the Happiness versus Sadness emotions.

Table 5 summarises the PPM classifier results for the three datasets. 76.0% accuracy was obtained for the Livejournal dataset, 80% accuracy for Aman's dataset, whereas 76.7% accuracy was obtained for Alm's dataset.

Table 5. PPM classification results for Happiness versus Sadness emotions produced by the PPM classifier for the LiveJournal, Aman, and Alm datasets.

Dataset	Accuracy (%)	Precision	Recall	F-measure
LiveJournal	76.0%	0.81	0.59	0.68
Aman	80.0%	0.74	0.66	0.69
Alm	76.7%	0.76	0.73	0.74

5.3 Experiments with Ekman's Emotion Classes

In these experiments, the PPM method was applied to the three datasets in order to classify Ekman's basic emotions.

For the LiveJournal's dataset, text related to the six emotions was first extracted. The Anger text consisted of 562 blogs, there were 143 blogs for Disgust, 177 blogs for Fear, 3601 blogs for Happiness, 1164 blogs for Sadness, and 521 blogs for Surprise). For Aman's dataset, text related to each of the six Ekman emotions was extracted directly according to the blog annotations. Aman's dataset is much smaller than the LiveJournal dataset (the Anger text contained 160 blogs, Disgust text contained 170 blogs, Fear 110 blogs, Happiness 530 blogs, Sadness 170 blogs, and Surprise 107 blogs). Similarly, for Alm's dataset, the text for each emotion could be extracted directly, (the Anger-Disgust contained 210 blogs, Fear 160 blogs, Happiness 440 blogs, Sadness 260 blogs, and Surprise 110 blogs).

Previous compression experiments with English text [21] have shown that a PPM order 5 model with escape method D is effective and therefore this was used for the classification experiments. At first a static model was built from the training data for each class. Ten-fold cross validation was then applied to evaluate the classification of the text according to Ekman's six basic emotions for all three datasets using the static models.

Table 6. PPM classification results for Ekman's emotions for the LiveJournal, Aman, and Alm datasets.

Dataset	Accuracy	Precision	Recall	F-measure
LiveJournal	87.8%	0.69	0.27	0.39
Aman	84.9%	0.50	0.41	0.45
Alm	69.1%	0.26	0.23	0.24

Table 6 summarises the best results that were obtained for the three datasets in terms of accuracy, precision, recall and F-measure. For example, the average accuracy for the classification of Ekman's six basic emotions for the LiveJournal dataset was 87.8%, for Aman's dataset was 84.9%, whereas for Alm's dataset was 69.1%.

A comparison was also made between using the PPM classifier for Ekman's classes on the LiveJournal dataset using text with and without basic pre-processing steps applied to it for blogs. The purpose of this experiment was to determine if the presence of the punctuation and digits were important for the classification or not. The pre-processing involved removing all punctuation and digits from the dataset. Table 7 shows the comparison of the classification results for the two versions of the LiveJournal dataset. The results show that the results for the PPM classifier do not change much when using the raw LiveJournal text compared to when the text was pre-processed first by removing the punctuation and digits.

Table 7. PPM classification results for Ekman's emotions for the two versions of the LiveJournal dataset with and without punctuation and digits.

Dataset	Accuracy	Precision	Recall	F-measure
LiveJournal text with punctuation and digits (i.e. raw text)	87.5%	0.70	0.27	0.39
LiveJournal text without punctuation and digits (pre-processed text)	87.8%	0.69	0.27	0.39

The confusion matrix that resulted from applying PPM to classify emotions for the pre-processed text is presented in Table 8. The training class is shown in the leftmost column with the testing class shown in the topmost row. The number of correct classifications made are shown in bold font.

Table 8. Confusion matrix for the PPM classification of the six basic emotions for the LiveJournal blogs.

Training	Anger	Disgust	Fear	Happiness	Sadness	Surprise
Anger	**46**	0	1	483	30	2
Disgust	0	**20**	0	113	9	1
Fear	0	0	**17**	134	19	7
Happiness	13	2	4	**3464**	105	13
Sadness	2	0	1	842	**317**	2
Surprise	2	1	1	428	50	**39**

Further PPM classification experiments were conducted using two versions of the text (with and without punctuation and digits) for both Aman's and Alm's datasets. The results are shown in Tables 9 and 10. The results show that as for the LiveJournal dataset, the removal of the punctuation and digits from Aman's dataset improves the classification noticeably, with accuracy increasing from 84.9% to 85.3%, precision from 0.50 to 0.52, recall from 0.41 to 0.43 and F- measure from 0.45 to 0.47. For Alm's dataset, there is also some improvement in these measures with accuracy increasing from 69.1% to 69.2%, precision improved from 0.26 to 0.30, recall not changed, F-measure improved from 0.24 to 0.26.

Table 9. PPM classification results on Ekman's emotions for the two versions of Aman's Dataset with and without punctuation and digits.

Dataset	Accuracy	Precision	Recall	F-measure
Aman's dataset with punctuation and digits (i.e. raw text)	84.9%	0.50	0.41	0.45
Aman's text without punctuation and digits (pre-processed text)	85.3%	0.52	0.43	0.47

Table 10. PPM classification results on Ekman's emotions for the two versions of Alm's dataset with and without punctuation and digits.

Dataset	Accuracy	Precision	Recall	F-measure
Alm's dataset with punctuation and digits (i.e. raw text)	69.1%	0.26	0.23	0.24
Alm's dataset without punctuation and digits (pre-processed text)	69.2%	0.30	0.23	0.26

Chaffer and Inkpen [5] used various traditional classifiers (ZeroR, Naïve Bayes, J48, SMO) implemented using Weka on both Aman's and Alm's datasets. Ghazi et al. [9] used both a flat SMO classifier and a two-level SMO classifier on the same two datasets. However, it is important to note that direct comparison between these studies is difficult due to the different processing and data selection methods used in each case (Table 11).

Table 11. Comparing accuracy results for Ekman's emotions for the three datasets.

Method (and reference)	Accuracy		Alm
	LiveJournal	Aman	
PPM [*This paper*]	87.8%	84.9%	69.1%
ZeroR [5]		68.5%	36.9%
Naïve Bayes [5]		73.0%	54.9%
J48 [5]		71.4%	47.5%
SMO [5]		81.2%	61.9%
Flat SMO [9]		61.7%	57.4%
Two-level SMO [9]		65.5%	56.6%

6 Conclusion and Future Work

We have described how the Prediction by Partial Matching (PPM) text compression scheme can be applied to automatically recognising emotions in text. Experimental results show that our new method outperforms other traditional data mining methods at this task. The PPM-based method processes all the characters in the text and therefore does not require explicit feature extraction as opposed to the other research methods that rely on identifying words as features.

Experiments with the PPM-based classifier were performed on three datasets: the LiveJournal dataset, Alm's dataset and Aman's dataset. Binary classification to recognise either the Happiness or Sadness emotions was applied on the three datasets. The experiments on the LiveJournal dataset achieved 76.0% and on Aman's dataset achieved 80.0% accuracy, whereas our method achieved 76.7% on Alm's dataset. Another binary classification experiment on emotional versus non-emotional sentences was applied to Aman's dataset using the PPM, Naïve Bayes, and SMO classifiers. Although the accuracy result for PPM (65.6%) was less than for two other classifiers (Naïve Bayes 68.8%, and SMO 68.8%), the PPM method achieved the best results in terms of precision, recall and F-measure for all classifiers that were compared.

Experiments at recognising Ekman's basic emotions using the PPM-based classifier were also performed on the three datasets. Our experiment on the LiveJournal's dataset achieved 87.8% accuracy, on Aman's dataset 84.9% accuracy and on Alm's dataset 69.1% accuracy. This is a significant improvement over previously published results that relied on traditional word-based data mining methods on the same datasets. We also found variations on accuracy, precision, recall, and F-measure when these texts were pre-processed to remove punctuation characters and digits. For Aman's and Alm's datasets, all measures were increased when we removed punctuation and digits from text prior to classification, although for the LiveJournal dataset, there was very little variation in performance.

In future, it would be interesting to explore this method on different languages such as Arabic and Chinese rather than English by performing additional experiments.

References

1. Alm, C.O., Roth, D., Sproat, R.: Emotions from text: machine learning for text-based emotion prediction. In: Proceedings of the Conference on Human Language Technology and Empirical Methods in Natural Language Processing, pp. 579–586. Association for Computational Linguistics (2005)
2. Almahdawi, A., Teahan, W.J.: Emotion recognition in text using PPM. In: Bramer, M., Petridis, M. (eds.) SGAI 2017. LNCS (LNAI), vol. 10630, pp. 149–155. Springer, Cham (2017). https://doi.org/10.1007/978-3-319-71078-5_13
3. Aman, S., Szpakowicz, S.: Identifying expressions of emotion in text. In: Matoušek, V., Mautner, P. (eds.) TSD 2007. LNCS (LNAI), vol. 4629, pp. 196–205. Springer, Heidelberg (2007). https://doi.org/10.1007/978-3-540-74628-7_27
4. Aman, S., Szpakowicz, S.: Using Roget's thesaurus for fine-grained emotion recognition. In: IJCNLp, pp. 312–318. Citeseer (2008)

5. Chaffar, S., Inkpen, D.: Using a heterogeneous dataset for emotion analysis in text. In: Butz, C., Lingras, P. (eds.) AI 2011. LNCS (LNAI), vol. 6657, pp. 62–67. Springer, Heidelberg (2011). https://doi.org/10.1007/978-3-642-21043-3_8

6. Cleary, J., Witten, I.: Data compression using adaptive coding and partial string matching. IEEE Trans. Commun. **32**(4), 396–402 (1984)

7. Ekman, P.: An argument for basic emotions. Cognit. Emot. **6**(3–4), 169–200 (1992)

8. Ekman, P.: Facial expressions. In: Handbook of Cognition and Emotion, vol. 16, pp. 301–320 (1999)

9. Ghazi, D., Inkpen, D., Szpakowicz, S.: Hierarchical versus flat classification of emotions in text. In: Proceedings of the NAACL HLT 2010 Workshop on Computational Approaches to Analysis and Generation of Emotion in Text, pp. 140–146. Association for Computational Linguistics (2010)

10. Howard, P.G.: The design and analysis of efficient lossless data compression systems. Ph.D. thesis, Brown University (1993)

11. Keshtkar, F.: A computational approach to the analysis and generation of emotion in text. Ph.D. thesis, Université d'Ottawa/University of Ottawa (2011)

12. Mahoui, M., Teahan, W.J., Sekhar, A.K.T., Chilukuri, S.: Identification of gene function using prediction by partial matching (PPM) language models. In: Proceedings of the 17th ACM Conference on Information and Knowledge Management, pp. 779–786. ACM (2008)

13. Mihalcea, R., Liu, H.: A corpus-based approach to finding happiness. In: AAAI Spring Symposium: Computational Approaches to Analyzing Weblogs, pp. 139–144 (2006)

14. Mishne, G., et al.: Experiments with mood classification in blog posts. In: Proceedings of ACM SIGIR 2005 Workshop on Stylistic Analysis of Text for Information Access, vol. 19, pp. 321–327. Citeseer (2005)

15. Moffat, A.: Implementing the PPM data compression scheme. IEEE Trans. Commun. **38**(11), 1917–1921 (1990)

16. Neviarouskaya, A., Prendinger, H., Ishizuka, M.: Compositionality principle in recognition of fine-grained emotions from text. In: ICWSM (2009)

17. Neviarouskaya, A., Prendinger, H., Ishizuka, M.: AM: textual attitude analysis model. In: Proceedings of the NAACL HLT 2010 Workshop on Computational Approaches to Analysis and Generation of Emotion in Text, pp. 80–88. Association for Computational Linguistics (2010)

18. Ramiandrisoa, F., Mothe, J., Benamara, F., Moriceau, V.: IRIT at e-Risk (2018)

19. Shaheen, S., El-Hajj, W., Hajj, H., Elbassuoni, S.: Emotion recognition from text based on automatically generated rules. In: 2014 IEEE International Conference on Data Mining Workshop (ICDMW), pp. 383–392. IEEE (2014)

20. Strapparava, C., Mihalcea, R.: Semeval-2007 task 14: affective text. In: Proceedings of the 4th International Workshop on Semantic Evaluations, pp. 70–74. Association for Computational Linguistics (2007)

21. Teahan, W.J.: Modelling English text. Ph.D. thesis, University of Waikato (1998)

22. Teahan, W.J., Harper, D.J.: Using compression-based language models for text categorization. In: Croft, W.B., Lafferty, J. (eds.) Language Modeling for Information Retrieval. The Springer International Series on Information Retrieval, vol. 13. Springer, Dordrecht (2003). https://doi.org/10.1007/978-94-017-0171-6_7

23. Yang, H., Willis, A., De Roeck, A., Nuseibeh, B.: A hybrid model for automatic emotion recognition in suicide notes. Biomed. Inform. Insights **5**(Suppl. 1), 17 (2012)

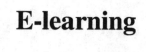

E-learning

Implementation of Augmented Reality in the Teaching of Young Children

Modafar Ati[1(⊠)], Hamis Abdullahi[2], Kamil Kabir[2],
and Masud Ahmed[2]

[1] Computer Science and Information Technology Department,
College of Engineering,
Abu Dhabi University, Abu Dhabi, United Arab Emirates
Modafar.ati@adu.ac.ae
[2] College of Engineering, Abu Dhabi University,
Abu Dhabi, United Arab Emirates
{1038178,1043012,1040506}@students.adu.ac.ae

Abstract. Learning to write can be exhausting for young children. In traditional teaching, children with different learning abilities are taught with the same rubric. This, in turn, impacts children that need extra attention to catch up with their peers, which leads them to suffer right from the early learning stages. Traditional teaching methods also are so rigid which makes them unable to automatically identify those children with less ability and in need of extra help. Hence, with the rapid development of ICT, innovative learning methods are sought to be important to allow children to be taught with different rubrics. The aim of this research is to improve the learning process for pre-school children via introducing Augmented Reality (AR) into the process, which, in turn, simplifies it as well as identifying children's abilities. The research introduces gamification to the process in order to ease the burden on children. Furthermore, we are trying to involve both the school as well the home to be part of the educational cycle so that parents are a part of the learning/educational process of their young children. Augmented reality combined with pleasing sound make the learning more interactive and enjoyable. The outcome of this research also helps parents to keep track of their children's learning. The paper also describes the deployment of the application in local schools as a pilot study so teachers can get feedback on students' learning curves and to fine tune the work further.

Keywords: Augmented reality · Education · Cloud computing

1 Introduction

The rapid evolvement of technology in recent years has provided numerous advancements in both the economy and society. Technology has provided many possibilities to the world of education as a whole, for example, looking at a complete heart beating in the middle of a classroom. It has already ceased to be science fiction and is entirely possible having the appropriate devices [1]. Augmented Reality (AR) is a term invented by a former Boeing researcher named Thomas P. Caudell. This was a

© Springer Nature Switzerland AG 2018
S. O. Al-mamory et al. (Eds.): NTICT 2018, CCIS 938, pp. 287–297, 2018.
https://doi.org/10.1007/978-3-030-01653-1_18

separate new direction of virtual reality not simply a new name for an existing concept, which propagated to become a distinctive branch or field on its own [1]. Unlike virtual reality, augmented reality unlocks and creates layers of digital information on top of the physical world that can be viewed through mobile devices. Teachers understand that learning expands, not just through reading and listening, but also through creating and interacting. Hence, with augmented reality [3], students can work in an interactive and fun way [2] to enhance their education by adding augmented objects into their real-life learning process.

1.1   Research Problem

The majority of parents are trying as much as possible to participate in their children's learning process. However, this participation does not occur, in most cases, at the early stages of child development and education. This is crucial because not only does literacy development start early in life and is highly correlated with school achievement, in limiting the child's experiences with language and literacy the more likely he or she will have difficulty in learning to read. Writing can be a complex exercise for young children, especially those that suffer from some sort of condition such as dysgraphia. Furthermore, the rapid development of ICT in recent years has enabled researchers to develop new methods that contribute to the education of children in a constructive manner.

School and home are separate entities that have different teaching methodologies in most cases. These differentiations have a negative impact on a child's learning. In order to minimize and eventually eliminate these differences, it is essential to align both methodologies to ensure that the learning process is consistent across both entities. Thus, a framework is required to increase the coupling and collaboration among the different entities associated with children's learning. The framework is needed not only to couple the school and home but also to decrease the burden and stress on the child's learning process.

1.2   Motivation

In order to reduce the learning stress on children while providing them with the traditional teaching methodologies, this research paper proposes a framework that integrates available technologies with traditional school paper-based learning to help literacy development. This will give the child the opportunity to learn at different bases depending on their learning ability. The framework proposed here is to create collaboration between school and parents where both parties can monitor development of the child 24/7. Hence, learning weaknesses can be identified as well as identifying conditions such as dysgraphia at an early stage because parents will be engaged in the monitoring of their children's development process.

Memorizing games are added to the process in order to make the learning more interactive and enjoyable. The purpose of this research is to help parents but mainly the child in developing an early relationship with one another by learning. Parents can acknowledge the skill level of their children at an early stage, children with dysgraphia or even dyslexia can be diagnosed, and measures can be put in place.

This paper is structured as related work, followed by the motivation that is led in this research. Section three, describes the framework of the proposed system. Results discussions are presented in section four. Conclusions are demonstrated in section five.

2 Related Work

It is well known that providing young children with exceptional writing experiences can lay good foundations for literacy learning at a very young age. As a result, this often increases the mental development of these young children. Although there is lack of in-depth research on writing, findings reveal that writing among young children that includes but is not limited to name writing is associated with later reading and other literacy skills [4]. In fact, in some cases, studies show that development in writing somehow predicts other literacy skills such as reading, spelling and decoding reading comprehension in first grade [5], and spelling in second grade [6].

At a young child level of writing, they often practise the pattern of alphabets one way or another, which leads to understanding the relationship between sounds and the letters they represent. Ultimately the child uses this to build on the understanding of the alphabetic principle that is a particular letter accounts for an individual word [7]. Some studies show that there is an actual evidence indicating that early writing is of great significant to the literacy development of young children. Therefore, it is imperative that writing should be integrated into young children's learning environment.

Every process in life requires many stages that lead to being developed fully. Young children's writing is not an exception and ample research indicates that young children progress through various stages of writing as they develop writing skills [7, 8].

In 2009 Schickedanz and Casbergue provided a comprehensive picture of preschoolers' writing development as they develop from drawing lines to meaningful letters. Young children often begin writing using small lines on pages that may not be of any resemblance to letters in the alphabet or drawing pictures that communicate or do not communicate a message. An article by Traci [9], shows that at this stage parents often encourage their young children by doing at least one of the following:

Air writing: where the child writes letters in the air first so as to strengthen the muscles and help the child progress faster,

Foamy fun: this occurs while taking a bath; the parent spatters a touch of shaving cream on the sidewall. Because of the cloud-like feeling of the cream, the child will associate writing with fun, and this will further his or her writing skills,

Learn your letters: The parent gives the child a size A4 paper and shows them different patterns of various letters.

Next, young children begin to make continuous scribbles while forming a con-sistent shape, often a looping pattern or zigzag. Then, writing starts to represent sep-arate letter-like symbols or forms. Another article [10] shows that to develop writing skills better at an early stage parents and teachers should make young children write their names first (name writing). Using their fingers when writing in sand trays, also strengthening the hand muscles, is an exciting tool for learning and unique writing experiences, and finally to trace drawings and keep a journal of previous work done [11].

After that stage, writing begins to look similar to actual letters of the alphabet and then children begin to associate shapes with letters. Consequently, children then start a technique called invented spelling which basically is when the child begins to spell words based on sounds heard in spoken language often beginning with isolated, conspicuous sounds and gradually moving toward accurate spelling [12, 13].

The ultimate objective of learning is to write a word accurately, however, it seems that there is inconsistency among young children. This is due to the fact that they keep going back and forth between these stages, usually as they move from a lower level of writing to a higher level, or other forms of writing (name writing, word writing, story writing). Sulzby and Teale [14] stated that all forms of writing are considered emergent and develop writing.

3 Proposed System

Integrating technology with traditional school learning can be established by creating a user ID for every child that enables both school and parents to use in order to monitor the development of the child 24/7. This can be done via generating a paper with the relevant alphabets at the first instance. This paper has QR code and an AR marker that are embedded within, as shown in Fig. 1. AR is used to identify the alphabets available on the sheet, while the QR has all the child's details such as id, name, sheet number, language used, etc. The framework was developed as a mobile application that pro-vides the child with a series of activities to perform such as writing manually on paper as shown in Fig. 3. This will then be uploaded to the server to be assessed based on the way the child performed the writing activity, with the help of character recognition algorithms that are used by the applications using OpenCv. The results will then be populated and made available on the child's account on the cloud. Both parents and teachers can view these results in order to check the learning progress. Furthermore, the school can check the progress of each child as well as the possibility of reviewing a group of children within the same class. The latter is based on each letter where the class average is calculated and then any weaknesses can be identified as shown in Figs. 6 and 7. In addition, the object recognition game embedded in the application is added as an incentive to encourage the child and also to enhance memorization skills as shown in Fig. 2.

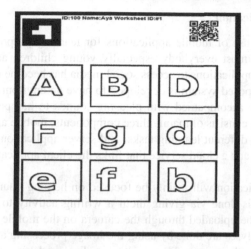

Fig. 1. Task worksheet prototype

Fig. 2. Game icon screen

Fig. 3. AR video of a child performing the writing activity

3.1   Design Specification

In recent times, the use of mobile applications for learning purposes and mostly for playing games by almost everybody especially young children and for most adults using mobile phone applications to access social media has become the major trend. On that account, the proposed system is developed to have an environment that comprises of augmented reality accompanied with pleasing sounds to hamper the learning process. The framework consists of mainly three components; mobile application, a printer used for printing the different learning tasks, and server application used for analyzing the completed tasks, and a cloud service that links the client application with that in the cloud.

The mobile application will mainly be focused on helping young children to learn how to write. This is done via giving them a writing activity to perform on paper, which, in turn, will be uploaded through the camera on the mobile device and later be analyzed. Image analyses are done by using the image processing algorithm embedded to the application in the server. The results are displayed via accessing the child's account available on the cloud for the parent or tutor to review and compliment on. Finally, to encourage the child to perform this activity, an inbuilt object recognition game has been added, whereby when the child plays it, they automatically get new activities to perform upon completion of each stage.

In general, the framework is designed as a mobile application embedded with an object recognition game. The application was developed using unity, while the cloud was developed using material design lite (mdl), for the server side. Also OpenCv was used for the image processing and using both AR toolkit and vuforia to select useful AR design and templates to be used for the children's interaction with the application. Each letter in the alphabet is divided at the pre-processing stage into a set of triangular meshes in order to capture the writing of the child. Figure 4 below shows one of the letter's image prior to commissioning the applications.

Finally, as stated above the main goal of this application is to assess the writing capability and performance of young children and to help them improve at their own speed without pressuring them to perform extra boring tasks they are not willing to do.

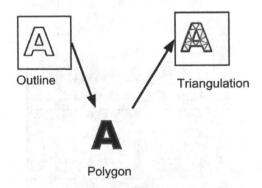

Fig. 4. Pre-processing of a letter image

3.2 Test Scenario

The test case involves prompting young children to use the object recognition game and use the work sheet to learn to write the alphabet. The framework assumes the child will write within the borders of the alphabet for it to be checked by the system. The overall scenario on how the framework works is as follows;

The parent of the child accesses the cloud in order to create an account for their child. Registration requires the user to provide a username and select an appropriate password. Following the completion of the child's registration, the parent can download and print a work sheet of an activity automatically generated by the system. The child, via using the mobile application, can start using the system by playing the object recognition game to enhance his/her memorization ability. Following completing the latter game, an AR embedded to the application would show a pop up video explaining to the child how to complete each given task that is identified by the AR marker on each activity, as shown in Fig. 3. Upon learning how to write the alphabet, the child is now required to complete the work sheet, and then take a snapshot of it. Upon taking the snapshot, the application will automatically recognize the user id and upload the sheet in to the user's workspace that is available on the cloud.

Upon uploading the completed sheet on the cloud, OpenCv code and image processing algorithms on the server will automatically analyze every new uploaded completed sheet and make the results available on demand to both parent and teachers.

In addition, when either parents or teachers access/login again, the system analyzes the progress of the child and presents a new sheet with a new alphabet as well as those letters that the child is having problems with and which he needs to practise more. Figure 5 below shows how the whole system works.

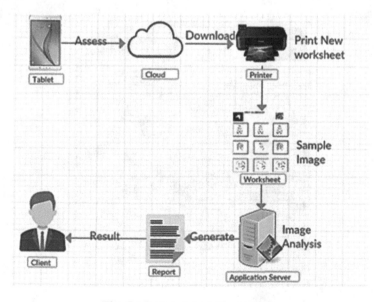

Fig. 5. System interaction scenario

4   Result and Discussions

A pilot study was taken in one of the local nurseries in Abu Dhabi to test how children feel about the application and how it will influence their learning process. So far, the mobile application has been developed specifically for educational purpose. As soon as a sheet is completed and a snapshot is taken, the server analyses the writing and provides a feedback of the progress of the child as a percentage. Figures 6 and 7 show the results of a typical sheet and the way that calculations are performed respectively for each individual letter. Added to that, the system also enables the school, with the right privileges, to check the progress of the whole class for a particular letter worked on by the children. The latter functionality, has the benefit of monitoring those letters most children are having difficulties with and provides a working solution to both parents and teachers. Figures 8 and 9 show the progress of a specific class for both upper and lower letter cases, which indicates clearly that students in that specific class are lacking skill in a given letter (e.g. W and w) which, in turn, gives an indication to the teacher to concentrate on these letters for forthcoming lessons.

A pilot study was conducted involving five children learning the letters (A, B, C) over two weeks. Table 1, shows the students and their percentage progress for the letter (A), while Figs. 10 and 11 show the individual progress for each student as per each letter. It can be seen that the children are progressing at their own pace without the need to increase the burden on them. The progress for letter (A) ranges between 14% and 31%. As a result, the parent can be part of the process in this case to encourage children to progress further.

Finally, the figures presented show a positive improvement in children's writing abilities.

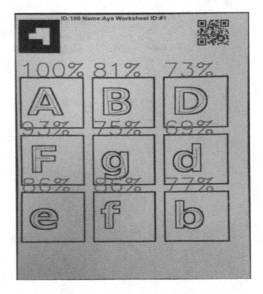

Fig. 6. Calculated worksheet

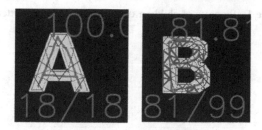

Fig. 7. Processing of letters

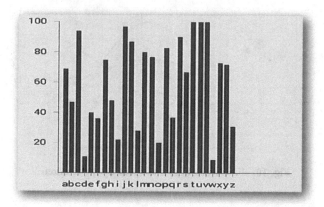

Fig. 8. Class learning progress per lower cases alphabet

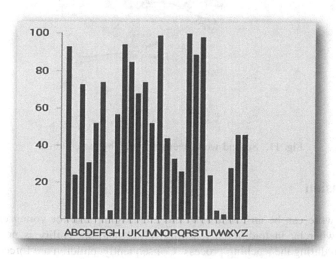

Fig. 9. Class learning progress per upper cases alphabet

Table 1. Progress of children over two week for the letter A

Name	Week one	Week two	Progress
Aya	40%	71%	31%
Ali	20%	43%	23%
Zayed	25%	39%	14%
Amna	33%	64%	31%
Mohamed	38%	60%	22%

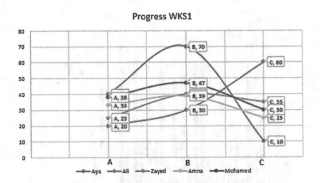

Fig. 10. First worksheet attempted by the children

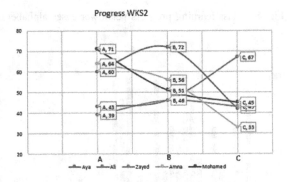

Fig. 11. Second worksheet attempted by the children

5 Conclusion

Learning to write can be an exhausting and challenging task for young children at a young age, whereby in traditional teaching, child learning ability is not taken into consideration during the teaching process. Consequently, children are forced to learn to write beyond their ability. The system built during this research demonstrated the improvement of learning that adds to the learning process of children at early stages. The results obtained show clear progress in children's ability to write the alphabet at an early age. The system can differentiate between students' ability and allows them to learn at their own pace. Furthermore, the system shows that parents are getting more involved in the monitoring as well as contributing to their children's progress. Finally, implementing such a system will ease the pressure on teaching institutes when implementing the proposed application, and also increase the participation of parents in the growth of their young children.

References

1. López, L.L.: eLearning Industry. https://elearningindustry.com/virtual-reality-augmented-reality-education. Accessed 24 Mar 2016
2. Nesloney, T.: edutopia. https://www.edutopia.org/blog/augmented-reality-new-dimensions-learning-drew-minock. Accessed 4 Nov 2013
3. Wu, H.-K., Lee, S.W.-Y., Chang, H.-Y., Liang, J.-C.: Current status, opportunities and challenges of augmented reality in education. Comput. Educ. **62**, 41–49 (2013)
4. Hammil, D.D.: What we know about correlates of reading. SAGE Except. Child. **70**(4), 453–469 (2004)
5. Shatil, E., Share, D.L., Levin, I.: On the contribution of kindergarten writing to grade 1 literacy: a longitudinal study in Hebrew. Appl. Psycholinguist. 1–21 (2000)
6. Dorit, A.: Continuity in children's literacy achievements: a longitudinal perspective from kindergarten to school. First Language **25**(3), 259–289 (2005)
7. Levin, I.A.D.: Perspectives on Language and Language Development, pp. 219–239 (2005)
8. Bloodgood, J.W.: What's in a name? Children's name writing and literacy acquisition. Read. Res. Q. **34**(3), 342–367 (1999)
9. Traci, G.: education.com. http://www.education.com/magazine/article/preschool-letter-writing/. Accessed 6 Dec 2013
10. Stewart, D.J.: TeachPreschool. https://www.teachpreschool.org/2014/05/03/6-ways-to-encourage-writing-in-preschool/. Accessed 3 May 2014
11. Gerde, H.K., Bingham, G.E., Wasik, B.A.: Writing in early childhood classrooms: guidance for best practices. Early Childhood Educ. J. **40**(6), 351–359 (2012)
12. Charles, R.: Children's categorization of speech sounds in English. National Council of Teachers of English, pp. 189–201 (1975)
13. Bear, D.R., Invernizzi, M., Templeton, S., Johnston, F.: Words Their Way: Word Study for Phonics, Vocabulary, and Spelling Instruction, p. 480. Person Education, London (2015)
14. Sulzby, E., Teale, W.H.: Writing development in early childhood. Educ. Horizons **64**(1), 8–12 (1985)

Author Index

Printed in the United States
By Bookmasters